ASE Test Preparation S

Automobile Test

Suspension and Steering (Test A4)

4th Edition

THOMSON

DELMAR LEARNING

Australia   Canada   Mexico   Singapore   Spain   United Kingdom   United States

THOMSON

DELMAR LEARNING

Thomson Delmar Learning's ASE Test Preparation Series
Automobile Test for Suspension and Steering (Test A4), 4th Edition

Vice President, Technology Professional Business Unit:
Gregory L. Clayton

Product Development Manager:
Kristen Davis

Product Manager:
Kim Blakey

Editorial Assistant:
Vanessa Carlson

Director of Marketing:
Beth A. Lutz

Marketing Specialist:
Brian McGrath

Marketing Coordinator:
Marissa Maiella

Production Manager:
Andrew Crouth

Production Editor:
Kara A. DiCaterino

Senior Project Editor:
Christopher Chien

XML Architect:
Jean Kaplansky

Cover Design:
Michael Egan

Cover Images:
Portion courtesy of DaimlerChrysler Corporation

ISBN: 1-4180-3881-4

NOTICE TO THE READER

Contents

Section 5 Sample Test for Practice

Section 6 Additional Test Questions for Practice

Section 7 Appendices

Preface

Delmar Learning is very pleased that you have chosen our ASE Test Preparation Series to prepare yourself for the automotive ASE Examination. These guides are available for all of the automotive areas including A1–A8, the L1 Advanced Diagnostic Certification, the P2 Parts Specialist, the C1 Service Consultant and the X1 Undercar Specialist. These guides are designed to introduce you to the Task List for the test you are preparing to take, give you an understanding of what you are expected to be able to do in each task, and take you through sample test questions formatted in the same way the ASE tests are structured.

If you have a basic working knowledge of the discipline you are testing for, you will find Delmar Learning's ASE Test Preparation Series to be an excellent way to understand the "must know" items to pass the test. These books are not textbooks. Their objective is to prepare the technician who has the requisite experience and schooling to challenge ASE testing. It cannot replace the hands-on experience or the theoretical knowledge required by ASE to master vehicle repair technology. If you are unable to understand more than a few of the questions and their explanations in this book, it could be that you require either more shop-floor experience or further study. Some resources that can assist you with further study are listed on the rear cover of this book.

Each book begins with an item-by-item overview of the ASE Task List with explanations of the minimum knowledge you must possess to answer questions related to the task. Following that there are 2 sets of sample questions followed by an answer key to each test and an explanation of the answers to each question. A few of the questions are not strictly ASE format but were included because they help teach a critical concept that will appear on the test. We suggest that you read the complete Task List Overview before taking the first sample test. After taking the first test, score yourself and read the explanation to any questions that you were not sure about, including the questions you answered correctly. Each test question has a reference back to the related task or tasks that it covers. This will help you to go back and read over any area of the task list that you are having trouble with. Once you are satisfied that you have all of your questions answered from the first sample test, take the additional tests and check them. If you pass these tests, you will be prepared to do well on the ASE test.

Our Commitment to Excellence

The 4th edition of Delmar Learning's ASE Test Preparation Series has been through a major revision with extensive updates to the ASE's task lists, test questions, and answers and explanations. Delmar Learning has sought out the best technicians in the country to help with the updating and revision of each of the books in the series.

About the Series Advisor

To promote consistency throughout the series, a series advisor took on the task of reading, editing, and helping each of our experts give each book the highest level of accuracy possible. Dan Perrin has served in the role of Series Advisor for the 4th edition of the ASE Test Preparation Series. Dan began ASE testing with the first series of tests in 1972 and has been continually certified ever since. He holds ASE master status in automotive, truck, collision, and machinist. He is also L1, L2, and alternated fuels certified, along with some others that have expired. He has been an automotive educator since 1979, having taught at the secondary, post-secondary, and industry levels. His service includes participation on boards that include the North American Council of Automotive Teachers (NACAT), the Automotive Industry Planning Council (AIPC), and the National Automotive Technicians Education Foundation (NATEF). Dan currently serves as the Executive Manager of NACAT and Director of the NACAT Education Foundation.

Thanks for choosing Delmar Learning's ASE Test Preparation Series. All of the writers, editors, Delmar Staff, and myself have worked very hard to make this series second to none. I know you are going to find this book accurate and easy to work with. It is our objective to constantly improve our product at Delmar by responding to feedback.

If you have any questions concerning the books in this series, you can email me at: autoexpert@trainingbay.com.

Dan Perrin
Series Advisor

The History and Purpose of ASE

ASE began as the National Institute for Automotive Service Excellence (NIASE). It was founded as a non-profit independent entity in 1972 by a group of industry leaders with the single goal of providing a means for consumers to distinguish between incompetent and competent technicians. It accomplishes this goal by testing and certification of repair and service professionals. From this beginning it has evolved to be known simply as ASE (Automotive Service Excellence) and today offers more than 40 certification exams in automotive, medium/heavy duty truck, collision, engine machinist, school bus, parts specialist, automobile service consultant, and other industry-related areas. At this time there are more than 400,000 professionals with current ASE certifications. These professionals are employed by new car and truck dealerships, independent garages, fleets, service stations, franchised service facilities, and more. ASE continues its mission by also providing information that helps consumers identify repair facilities that employ certified professionals through its Blue Seal of Excellence Recognition Program. Shops that have a minimum of 75% of their repair technicians ASE certified and meet other criteria can apply for and receive the Blue Seal of Excellence Recognition from ASE.

ASE recognized that educational programs serving the service and repair industry also needed a way to be recognized as having the faculty, facilities, and equipment to provide a quality education to students wanting to become service professionals. Through the combined efforts of ASE, industry, and education leaders, the non-profit National Automotive Technicians Education Foundation (NATEF) was created to evaluate and recognize training programs. Today more than 2000 programs are ASE certified under the standards set by the service industry. ASE/NATEF also has a certification of industry (factory) training program known as CASE. CASE stands for Continuing Automotive Service Education and recognizes training provided by replacement parts manufacturers as well as vehicle manufacturers.

ASE certification testing is administered by the American College Testing (ACT). Strict standards of security and supervision at the test centers insure that the technician who holds the certification earned it. Additionally ASE certification also requires that the person passing the test to be able to demonstrate that they have two years of work experience in the field before they can be certified. Test questions are developed by industry experts that are actually working in the field being tested. There is more detail on how the test is developed and administered in the next section. Paper and pencil tests are administered twice a year at over seven hundred locations in the United States. Computer based testing is now also available with the benefit of instant test results at certain established test centers. The certification is valid for five years and can be recertified by retesting. So that consumers can recognize certified technicians, ASE issues a jacket patch, certificate, and wallet card to certified technicians and makes signs available to facilities that employ ASE certified technicians.

You can contact ASE at any of the following:

National Institute for Automotive Service Excellence
101 Blue Seal Drive S.E.
Suite 101
Leesburg, VA 20175
Telephone 703-669-6600
FAX 703-669-6123
www.ase.com

WE SUPPORT
PROFESSIONAL CERTIFICATION
THROUGH THE
National Institute for
AUTOMOTIVE
SERVICE
EXCELLENCE

Take and Pass Every ASE Test

Participating in an Automotive Service Excellence (ASE) voluntary certification program gives you a chance to show your customers that you have the "know-how" needed to work on today's modern vehicles. The ASE certification tests allow you to compare your skills and knowledge to the automotive service industry's standards for each specialty area.

If you are the "average" automotive technician taking this test, you are in your mid-thirties and have not attended school for about fifteen years. That means you probably have not taken a test in many years. Some of you, on the other hand, have attended college or taken postsecondary education courses and may be more familiar with taking tests and with test-taking strategies. There is, however, a difference in the ASE test you are preparing to take and the educational tests you may be accustomed to.

How are the tests administered?

ASE test are administered at over 750 test sites in local communities. Paper and pencil tests are the type most widely available to technicians. Each tester is given a booklet containing questions with charts and diagrams where required. You can mark in this test booklet but no information entered in the booklet is scored. Answers are recorded on a separate answer sheet. You will enter your answers, using a number 2 pencil only. ASE recommends you bring four sharpened number 2 pencils that have erasers. Answer choices are recorded by coloring in the blocks on the answer sheet. The answer sheets are scanned electronically and the answers tabulated. For test security, test booklets include randomly generated questions. Your answer key must be matched to the proper booklet so it is important to correctly enter the booklet serial number on the answer sheet. All instructions are printed on the test materials and should be followed carefully.

ASE has introduced Computer Based Testing (CBT) at some locations. While the test content is the same for both testing methods the CBT tests have some unique requirements and advantages. It is strongly recommended that technicians considering the CBT tests go the ASE web page at www.ASE.com and review the conditions and requirements for this type of test. There is a demonstration of a CBT that allows you to experience this type of test before you register. Some technicians find this style of testing provides an advantage, while others find operating the computer a distraction. One significant benefit of CBT is the availability of instant results. You can receive your test results before you leave the test center. CBT testing also offers increased flexibility in scheduling. The cost for taking CBTs is slightly higher than paper and pencil tests and the number of testing sites is limited. The first time test taker may be more comfortable with the paper and pencil tests but technicians now have a choice.

Who Writes the Questions?

The questions are written by service industry experts in the area being tested. Each area will have its own technical experts. Questions are entirely job related. They are designed to test the skills you need to be a successful technician. Theoretical knowledge is important and necessary to answer the questions, but the ability to apply that knowledge is the basis of ASE test questions.

Each question has its roots in an ASE "item-writing" workshop where service representatives from automobile manufacturers (domestic and import), aftermarket parts and equipment manufacturers, working technicians, and vocational educators meet in a workshop setting to share ideas and translate them into test questions. Each test question written by these experts must survive review by all members of the group.

The questions are written to deal with practical application of soft skills and system knowledge experienced by technicians in their day-to-day work.

All questions are pre-tested and quality-checked on a national sample of technicians. Those questions that meet ASE standards of quality and accuracy are included in the scored sections of the tests; the "rejects" are sent back to the drawing board or discarded altogether.

Each certification test is made up of between forty and eighty multiple-choice questions.

Note: Each test could contain additional questions that are included for statistical research purposes only. Your answers to these questions will not affect your score, but since you do not know which ones they are, you should answer all questions on the test. The five-year Recertification Test will cover the same content areas as those listed above. However, the number of questions in each content area of the Recertification Test will be reduced by about one-half.

Objective Tests

A test is called an objective test if the same standards and conditions apply to everyone taking the test and there is only one correct answer to each question.

Objective tests primarily measure your ability to recall information. A well-designed objective test can also test your ability to understand, analyze, interpret, and apply your knowledge. Objective tests include true-false, multiple choice, fill in the blank, and matching questions. ASE's tests consist exclusively of four-part multiple-choice objective questions.

The following are some strategies that may be applied to your tests.

Before beginning to take an objective test, quickly look over the test to determine the number of questions, but do not try to read through all of the questions. In an ASE test, there are usually between forty and eighty questions, depending on the subject. Read through each question before marking your answer. Answer the questions in the order they appear on the test. Leave the questions blank that you are not sure of and move on to the next question. You can return to those unanswered questions after you have finished the others. They may be easier to answer at a later time after your mind has had additional time to consider them on a subconscious level. In addition, you might find information in other questions that will help you recall the answers to some of them.

Do not be obsessed by the apparent pattern of responses. For example, do not be influenced by a pattern like **D, C, B, A, D, C, B, A** on an ASE test.

There is also a lot of folk wisdom about taking objective tests. For example, there are those who would advise you to avoid response options that use certain words such as *all, none, always, never, must,* and *only,* to name a few. This, they claim, is because nothing in life is exclusive. They would advise you to choose response options that use words that allow for some exception, such as *sometimes, frequently, rarely, often, usually, seldom,* and *normally.* They would also advise you to avoid the first and last option (A and D) because test writers, they feel, are more comfortable if they put the correct answer in the middle (B and C) of the choices. Another recommendation often offered is to select the option that is either shorter or longer than the other three choices because it is more likely to be correct. Some would advise you to never change an answer since your first intuition is usually correct.

Although there may be a grain of truth in this folk wisdom, ASE test writers try to avoid them and so should you. There are just as many **A** answers as there are **B** answers, just as many **D** answers as **C** answers. As a matter of fact, ASE tries to balance the answers at about 25 percent per choice **A, B, C,** and **D.** There is no intention to use "tricky" words, such as outlined above. Put no credence in the opposing words "sometimes" and "never," for example.

Multiple-choice tests are sometimes challenging because there are often several choices that may seem possible, and it may be difficult to decide on the correct choice. The best strategy, in this case, is to first determine the correct answer before looking at the options. If you see the answer you decided on, you should still examine the options to make sure that none seem more correct than yours. If you do not know or are not sure of the answer, read each option very carefully and try to eliminate those options that you know to be wrong. That way, you can often arrive at the correct choice through a process of elimination.

If you have gone through all of the test and you still do not know the answer to some of the questions, then guess. Yes, guess. You then have at least a 25 percent chance of being correct. If you leave the question blank, you have no chance. Your score is based on the number of questions answered correctly.

Preparing for the Exam

The main reason we have included so many sample and practice questions in this guide is, simply, to help you learn what you know and what you don't know. We recommend that you work your way through each question in this book. Before doing this, carefully look through Section 3; it contains a description and explanation of the question types you'll find on an ASE exam.

Once you understand what the questions will look like, move to the sample test. Answer one of the sample questions (Section 5) then read the explanation (Section 7) to the answer for that question. If you don't feel you understand the reasoning for the correct answer, go back and read the overview (Section 4) for the task that is related to that question. If you still don't feel you have a solid understanding of the material, identify a good source of information on the topic, such as a textbook, and do some more studying.

After you have completed all of the sample test items and reviewed your answers, move to the additional questions (Section 6). This time answer the questions as if you were taking an actual test. Do not use any reference or allow any interruptions in order to get a feel for how you will do on an actual test. Once you have answered all of the questions, grade your results using the answer key in Section 7. For every question that you gave a wrong answer to, study the explanations to the answers and/or the overview of the related task areas. Try to determine the root cause for your missing the question. The easiest thing to correct is learning the correct technical content. The hardest thing to correct is behaviors that lead you to a wrong conclusion. If you knew the information but still got it wrong there is a behavior problem that will need to be corrected. An example would be reading too quickly and skipping over words that affect your reasoning. If you can identify what you did that caused you to answer the question incorrectly you can eliminate that cause and improve your score. Here are some basic guidelines to follow while preparing for the exam:

- Focus your studies on those areas you are weak in.

- Be honest with yourself while determining if you understand something.

- Study often but in short periods of time.

- Remove yourself from all distractions while studying.

- Keep in mind the goal of studying is not just to pass the exam, the real goal is to learn!

- Prepare physically by getting a good night's rest before the test and eat meals that provide energy but do not cause discomfort.

- Arrive early to the test site to avoid long waits as test candidates check in and to allow all of the time available for your tests.

During the Test

On paper and pencil tests you will be placing your answers on a sheet where you will be required to color in your answer choice. Stray marks or incomplete erasures may be picked up as an answer by the electronic reader, so be sure only your answers end up on the sheet. One of the biggest problems an adult faces in test taking, it seems, is placing the answer in the correct spot on the answer sheet. Make certain that you mark your answer for, say, question 21, in the space on the answer sheet designated for the answer for question 21. A correct response in the wrong line will probably result in two questions being marked wrong, one with two answers (which could include a correct answer but will be scored wrong) and the other with no answer. Remember, the answer sheet on the written test is machine scored and can only "read" what you have colored in.

If you finish answering all of the questions on a test and have remaining time, go back and review the answers to those questions that you were not sure of. You can often catch careless errors by using the remaining time to review your answers. Carefully check your answer sheet for blank answer blocks or missing information.

At practically every test, some technicians will invariably finish ahead of time and turn their papers in long before the final call. Some technicians may be doing recertification tests and others may be taking fewer tests than you. Do not let them distract or intimidate you.

It is not wise to use less than the total amount of time that you are allotted for a test. If there are any doubts, take the time for review. Any product can usually be made better with some additional effort. A test is no exception. It is not necessary to turn in your test paper until you are told to do so.

Testing Time Length

An ASE written test session is four hours. You may attempt from one to a maximum of four tests in one session. It is recommended, however, that no more than a total of 225 questions be attempted at any test session. This will allow for just over one minute for each question.

Visitors are not permitted at any time. If you wish to leave the test room, for any reason, you must first ask permission. If you finish your test early and wish to leave, you are permitted to do so only during specified dismissal periods.

You should monitor your progress and set an arbitrary limit to how much time you will need for each question. This should be based on the number of questions you are attempting. It is suggested that you wear a watch because some facilities may not have a clock visible to all areas of the room.

Computer-Based Tests are allotted a testing time according to the number of questions ranging from one half hour to one and one half hours. Advanced level tests are allowed two hours. This time is by appointment and you should be sure to be on time to insure that you have all of the time allocated. If you arrive late for a CBT test appointment you will only have the amount of time remaining on your appointment.

Your Test Results!

You can gain a better perspective about tests if you know and understand how they are scored. ASE's tests are scored by American College Testing (ACT), a nonpartial, unbiased organization having no vested interest in ASE or in the automotive industry.

Each question carries the same weight as any other question. For example, if there are fifty questions, each is worth 2 percent of the total score. The passing grade is 70 percent. That means you must correctly answer thirty-five of the fifty questions to pass the test.

The test results can tell you:

- where your knowledge equals or exceeds that needed for competent performance, or

- where you might need more preparation.

Your ASE test score report is divided into content areas and will show the number of questions in each content area and how many of your answers were correct. These numbers provide information about your performance in each area of the test. However, because there may be a different number of questions in each content area of the test, a high percentage of correct answers in an area with few questions may not offset a low percentage in an area with many questions.

It should be noted that one does not "fail" an ASE test. The technician who does not pass is simply told "More Preparation Needed." Though large differences in percentages may indicate problem areas, it is important to consider how many questions were asked in each area. Since each test evaluates all phases of the work involved in a service specialty, you should be prepared in each area. A low score in one area could keep you from passing an entire test.

There is no such thing as average. You cannot determine your overall test score by adding the percentages given for each task area and dividing by the number of areas. It doesn't work that way because there generally are not the same number of questions in each task area. A task area with twenty questions, for example, counts more toward your total score than a task area with ten questions.

Your test report should give you a good picture of your results and a better understanding of your strengths and weaknesses for each task area.

If you fail to pass the test, you may take it again at any time it is scheduled to be administered. You are the only one who will receive your test score. Test scores will not be given over the telephone by ASE nor will they be released to anyone without your written permission.

Types of Questions on an ASE Exam

ASE certification tests are often thought of as being tricky. They may seem to be tricky if you do not completely understand what is being asked. The following examples will help you recognize certain types of ASE questions and avoid common errors.

Paper-and-pencil tests and computer-based test questions are identical in content and difficulty. Most initial certification tests are made up of forty to eighty multiple-choice questions. Multiple-choice questions are an efficient way to test knowledge. To answer them correctly, you must think about each choice as a possibility, and then choose the one that best answers the question. To do this, read each word of the question carefully. Do not assume you know what the question is about until you have finished reading it.

About 10 percent of the questions on an actual ASE exam will use an illustration. These drawings contain the information needed to correctly answer the question. The illustration must be studied carefully before attempting to answer the question. Often, technicians look at the possible answers then try to match up the answers with the drawing. Always do the opposite; match the drawing to the answers. When the illustration is showing an electrical schematic or another system in detail, look over the system and try to figure out how the system works before you look at the question and the possible answers.

Multiple-Choice Questions

The most common type of question used on ASE Tests is the multiple-choice question. This type of question contains three "distracters" (wrong answers) and one "key" (correct answer). When the questions are written effort is made to make the distracters plausible to draw an inexperienced technician to one of them. This type of question gives a clear indication of the technician's knowledge. Using multiple criteria including cross-sections by age, race, and other background information, ASE is able to guarantee that a question does not bias for or against any particular group. A question that shows bias toward any particular group is discarded. If you encounter a question that you are unsure of, reverse engineer it by eliminating the items that it cannot be. For example:

A rocker panel is a structural member of which vehicle construction type?

A. Front-wheel drive
B. Pickup truck
C. Unibody
D. Full-frame

Analysis:

This question asks for a specific answer. By carefully reading the question, you will find that it asks for a construction type that uses the rocker panel as a structural part of the vehicle.

Answer A is wrong. Front-wheel drive is not a vehicle construction type.
Answer B is wrong. A pickup truck is not a type of vehicle construction.
Answer C is correct. Unibody design creates structural integrity by

welding parts together, such as the rocker panels, but does not require exterior cosmetic panels installed for full strength.
Therefore, the correct answer is C. If the question was read quickly and the words "construction type" were passed over, answer A may have been selected.
Answer D is wrong. Full-frame describes a body-over-frame construction type that relies on the frame assembly for structural integrity.
Therefore, the correct answer is C. If the question was read quickly and the words "construction type" were passed over, answer A may have been selected.

ECEPT Questions

Another type of question used on ASE tests has answers that are all correct except one. The correct answer for this type of question is the answer that is wrong. The word "**EXCEPT**" will always be in capital letters. You must identify which of the choices is the wrong answer. If you read quickly through the question, you may overlook what the question is asking and answer the question with the first correct statement. This will make your answer wrong. An example of this type of question and the analysis is as follows:

All of the following are tools for the analysis of structural damage **EXCEPT**:

A. height gauge
B. tape measure.
C. dial indicator.
D. tram gauge.

Analysis:

The question really requires you to identify the tool that is not used for analyzing structural damage. All tools given in the choices are used for analyzing structural damage except one. This question presents two basic problems for the test-taker who reads through the question too quickly. It may be possible to read over the word "**EXCEPT**" in the question or not think about which type of damage analysis would use answer C. In either case, the correct answer may not be selected. To correctly answer this question, you should know what tools are used for the analysis of structural damage. If you cannot immediately recognize the incorrect tool, you should be able to identify it by analyzing the other choices.

Answer A is wrong. A height gauge may be used to analyze structural damage.
Answer B is wrong. A tape measure may be used to analyze structural damage.
Answer C is correct. A dial indicator may be used as a damage analysis tool for moving parts, such as wheels, wheel hubs, and axle shafts, but would not be used to measure structural damage.
Answer D is wrong. A tram gauge is used to measure structural damage.

Technician A, Technician B Questions

The type of question that is most popularly associated with an ASE test is the "Technician A says . . . Technician B says . . . Who is right?" type. In this type of question, you must identify the correct statement or statements. To answer this type of question correctly, you must carefully read each technician's statement and judge it on its own merit to determine if the statement is true.
Sometimes this type of question begins with a statement about some analysis or repair procedure. This is often referred to as the stem of the question and provides the setup or background information required to understand the conditions the question is based on. This is followed by two statements about the cause of the concern, proper inspection, identification, or repair choices. You are asked whether the first statement, the second statement, both statements, or neither statement is correct.

Analyzing this type of question is a little easier than the other types because there are only two ideas to consider although there are still four choices for an answer.

Technician A, Technician B questions are really double true or false questions. The best way to analyze this kind of question is to consider each technician's statement separately. Ask yourself, is A true or false? Is B true or false? Then select your answer from the four choices. An important point to remember is that an ASE Technician A, Technician B question will never have Technician A and B directly disagreeing with each other. That is why you must evaluate each statement independently.

An example of this type of question and the analysis of it follows.

A vehicle comes into the shop with a gas gauge that will not register above one half full. When the sending unit circuit is disconnected the gauge reads empty and when it is connected to ground the gauge goes to full. Technician A says that the sending unit is shorted to ground. Technician B says the gauge circuit is working and the sending unit is likely the problem. Who is right?

A. A only
B. B only
C. Both A and B
D. Neither A nor B

Analysis:

Reading of the stem of the question sets the conditions of the customer concern and establishes what information is gained from testing. General knowledge of gauge circuits and test procedures are needed to correctly evaluate the technician's conclusions. Note: Avoid being distracted by experience with unusual or problem vehicles that you may have worked on, Other technicians taking the same test do not have that knowledge, so it should not be used as the basis of your answers.

Technician A is wrong because a shorted to ground sending unit would produce a gauge reading equivalent to the test conditions of a grounding the circuit and produce a full reading.
Technician B is correct because the gauge spans when going from an open circuit to a completely
grounded circuit. This would tend to indicate that the problem had to be in the sending unit.
Answer C is not correct. Both technicians are identifying the problem as a sending unit but technician A qualified the problem as a specific type of failure (grounded) that would not have caused the symptoms of the vehicle.
Answer D is not correct because technician B's diagnosis is a possible cause of the conditions identified.

Most-Likely Questions

Most-Likely questions are somewhat difficult because only one choice is correct while the other three choices are nearly correct. An example of a Most-Likely-cause question is as follows:

The Most-Likely cause of reduced turbocharger boost pressure may be a:

A. wastegate valve stuck closed.
B. wastegate valve stuck open.
C. leaking wastegate diaphragm.
D. disconnected wastegate linkage.

Analysis:

Answer A is wrong. A wastegate valve stuck closed increases turbocharger boost pressure.

Answer B is correct. A wastegate valve stuck open decreases turbocharger boost pressure.
Answer C is wrong. A leaking wastegate valve diaphragm increases turbocharger boost pressure.
Answer D is wrong. A disconnected wastegate valve linkage will increase turbocharger boost pressure.

LEAST-Likely Questions

Notice that in Most-Likely questions there is no capitalization. This is not so with LEAST-Likely type questions. For this type of question, look for the choice that would be the LEAST-Likely cause of the described situation. Read the entire question carefully before choosing your answer. An example is as follows:

What is the LEAST-Likely cause of a bent pushrod?

A. Excessive engine speed
B. A sticking valve
C. Excessive valve guide clearance
D. A worn rocker arm stud

Analysis:

Answer A is wrong. Excessive engine speed may cause a bent pushrod.
Answer B is wrong. A sticking valve may cause a bent pushrod.
Answer C is correct. Excessive valve clearance will not generally cause a bent pushrod.
Answer D is wrong. A worn rocker arm stud may cause a bent pushrod.

You should avoid relating questions to those unusual situations that you may have encountered and answer based on the technical and mechanical possibilities.

Summary

There are no four-part multiple-choice ASE questions having "none of the above" or "all of the above" choices. ASE does not use other types of questions, such as fill-in-the-blank, completion, true-false, word-matching, or essay. ASE does not require you to draw diagrams or sketches. If a formula or chart is required to answer a question, it is provided for you. There are no ASE questions that require you to use a pocket calculator.

4. Overview of the Task List

Suspension and Steering (Test A4)

The following section includes the task areas and task lists for this test and a written overview of the topics covered in the test.

The task list describes the actual work you should be able to do as a technician that you will be tested on by the ASE. This is your key to the test and you should review this section carefully. We have based our sample test and additional questions upon these tasks. The overview section will also support your understanding of the task list. ASE advises that the questions on the test may not equal the number of tasks listed; the task lists tell you what ASE expects you to know how to do and be ready to be tested upon.

At the end of each question in the Sample Test and Additional Test Questions sections, a letter and number will be used as a reference back to this section for additional study. Note the following example: **A.3.1**

A. Steering Systems Diagnosis and Repair (10 Questions)

Task A.3 **Steering Linkage (3 Questions)**

Task A.3.1 **Inspect and adjust (where applicable) front and rear steering linkage geometry (including parallelism and vehicle ride height)**

Example:

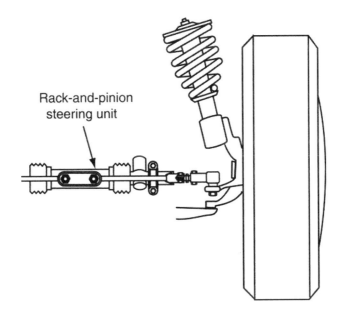

Rack-and-pinion steering unit

1. An inspection is being performed on the rack and pinion steering system shown in the figure. All of the following should be checked **EXCEPT:**

A. the ball joints.
B. the tires.
C. the Pitman arm.
D. the tie-rods.

(A.3.1)

Analysis:

Question #1
Answer A is wrong. Ball joints should be inspected within the steering system for wear and looseness.
Answer B is wrong. Tires should be inspected along with the steering system for abnormal wear.
Answer C is correct. A rack and pinion steering system does not have a Pitman arm.
Answer D is wrong. Tie-rods should be inspected as part of the steering system for wear and looseness.

Task List and Overview

A. Steering Systems Diagnosis and Repair (10 Questions)

Task A.1   **Steering Columns (3 Questions)**

Task A.1.1   **Diagnose steering column noises and steering effort concerns (including manual and electronic tilt and telescoping mechanisms); determine needed repairs.**

The steering column is the direct connection from the steering wheel to the steering gear. Although there are many designs, some basic characteristics apply to most columns. On all new vehicles, there is an air bag located in the center of the steering wheel. Most vehicles will incorporate (in the steering wheel and column) functions such as turn signals, headlights, hazard lights, ignition lock, horn, ignition switch, wipers, windshield washers, and cruise control switches. Some designs will also have additional features such as radio volume and station switching. A wiring harness runs alongside the column.

To add to driver comfort, many steering columns tilt and may even telescope up and down.

Diagnosing steering complaints always begins with a complete inspection of the **steering column**. Worn steering column bearings or loose mounts can cause noise and looseness in the steering column. Check for excessive up-and-down and side-to-side movement in the steering wheel and column. A broken support plate or other worn internal column parts can cause looseness or binding in the steering column.

In a **tilt steering column,** worn or loose pivots that allow the column to tilt will result in excessive steering wheel movement. In many cases, it will be necessary to remove the steering wheel and disassemble the steering column to properly identify the worn parts.

Task A.1.2   **Inspect and replace steering column, steering shaft U-joint(s), flexible coupling(s), collapsible columns, and steering wheels (includes steering wheels and columns equipped with air bags and/or other steering wheel/column mounted controls, sensors, and components).**

Before removing the steering column, point the front wheels in the straight-ahead position and make sure the column is in the lock position. This will prevent damage to the air bag clock spring (spiral cable) and will help line-up the steering column with the steering gear during reassembly. For vehicles equipped with air bags, disable the air bag system following the manufacturer's procedure. Usually, this involves removing the battery cables or air bag fuse.

After removing the horn pad (and/or air bag), remove the steering wheel retaining nut and check alignment marks on the steering wheel and shaft. If no marks are found, scribe marks to ensure proper reinstallation of the steering wheel. Using the proper puller, remove the steering wheel. Never use a hammer, slide-hammer, or knock-off puller to remove the steering wheel. Damage to the column or column bearings can result.

Disconnect the wiring harness, bottom steering coupler retaining bolt, and transmission linkage from the column. Mark the position of the steering column coupler to the steering gear. Remove, if necessary, any dash trim to gain access to the steering column mount under the dashboard. Carefully remove the steering column from the vehicle.

Inspect all external components including the flex coupler and lower column U-joint (if equipped). Many vehicles are designed with an intermediate shaft containing two U-joints. Inspect the U-joints for wear, looseness and binding. Disassembly of the steering column will be necessary to inspect internal components of the column.

Task A.1.3 Disarm, enable, and properly handle airbag system components during vehicle service following manufacturers' procedures.

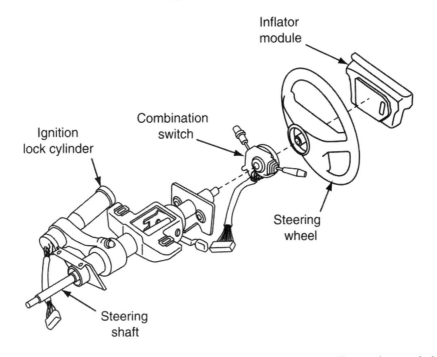

The supplemental inflatable restraint system (air bag) must be disarmed properly before any work is performed on the steering column. Failure to disarm the air bag may result in the accidental deployment of the air bag, which can cause personal injury and unnecessary repairs to the air bag system. Always consult the service manual for recommended procedures on disabling the air bag system.

The following is a list of basic procedures and precautions that should be followed when working with and around air bag systems.

- Always disable the air bag system before servicing the air bag or any components near or around the steering column and/or air bag system.

- Never subject the inflator module to temperatures greater than 175°F (79.4°C).

- If any air bag component is dropped, it should be replaced.

- Never test any air bag component with electrical test equipment unless instructed to do so by the factory service manual.

- When carrying a live (not deployed) air bag, always point the bag and trim cover away from you.

- When placing a live air bag on a workbench, always face the air bag and trim cover up.

- Discarded live air bags must be deployed. Follow manufacturer's procedure for proper disposal.

- Lock the steering wheel in place whenever removing the steering wheel, steering column, or steering gear to prevent damage to the clock spring. The clock spring maintains a continuous electrical connection as the steering wheel rotates between the inflator module and air bag controller.

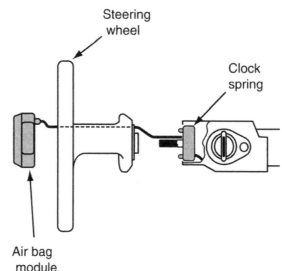

Task A.2   Steering Units (4 Questions)

Task A.2.1   Diagnose steering gear (non-rack and pinion type) noises, binding, vibration, free play, steering effort, steering pull (lead), and leakage concerns; determine needed repairs.

Conventional steering gears utilize the **recirculating ball design.** Many conventional steering gear problems are due to internal wear. Start your steering inspection by turning the steering wheel full left and full right, noting any noise, binding, roughness, looseness, or need for excessive effort. Check lubrication level. Inspect the steering gear for indications of fluid leakage from the gear and at hose connections for power steering units. Loss of fluid may cause increased steering effort and erratic steering. Fluid loss on power steering units may cause a whine or growling noise. A loose or worn power steering belt may cause a squeal while turning the steering wheel, and may also cause erratic steering and increased steering effort. On power steering systems, a missing belt will result in loss of power steering assist.

A worn sector shaft, worm shaft, or bearings can cause binding and roughness while turning the steering wheel. Excessively worn steering gears will have to be either overhauled or replaced.

Inspect the steering gear mounting bolts. Loose gear-to-frame mounting bolts will cause excessive free play, wandering, and may even cause a vibration. Inspect the frame where the steering gear is mounted for corrosion and cracks. Rust can weaken the frame and cause the gear to flex or pull away causing a potentially dangerous situation.

Task A.2.2   Diagnose rack and pinion steering gear noises, binding, vibration, free play, steering effort, steering pull (lead), and leakage concerns; determine needed repairs.

Most new vehicles are equipped with a **rack and pinion** steering gear. The rack and pinion steering gear is a much simpler design. The steering column is connected directly to the sector shaft, called the pinion. The pinion operates the rack assembly, which moves the tie-rods and the steering knuckles. On power steering rack and pinion gears, a control valve directs the flow of fluid from one side of the rack to the other.

Diagnose problems with a rack and pinion steering gear following the same basic procedures as with the conventional steering gear. Turn the steering wheel full left and full right, noting any noise, binding, roughness, looseness, or need for excessive effort. Inspect the mounting brackets and bushings for wear, which can cause loose or wandering steering. A binding or roughness while turning the wheel may indicate a worn rack gear. A worn steering rack may also cause excessive steering effort. Carefully inspect the rack for leaks at the bellows boots and at the pinion seal.

Task A.2.3 Inspect power steering fluid level and condition; determine fluid type and adjust fluid level in accordance with vehicle manufacturers' recommendations.

Power steering fluid reservoirs are either integrated with the pump or remotely located. Many remote reservoirs are transparent and have markings on the outside to indicate the fluid level. Remote reservoirs are usually marked: **FULL COLD, FULL HOT**, or **MIN, MAX**. If the fluid is checked with a dipstick, wipe the cap and area clean to prevent dirt from entering the system.

Check the fluid level at normal operating temperature (approximately 175°F.). Make sure the fluid level is at least MIN or COLD before running the engine to prevent damage to the pump or gear. A low fluid level may indicate a leak.

When checking the fluid level, note the fluid condition. Discolored fluid or signs of particles may be an indication of wear to the pump, hoses, or gear. Foamy fluid is caused by air in the system.

Before adding fluid, check with the factory manual for the proper fluid.

Task A.2.4 Inspect, adjust, align, and replace power steering pump belt(s) and tensioners.

The power steering pump is driven by either a conventional V-belt or serpentine belt. Most new vehicles are equipped with serpentine belts that incorporate automatic tensioners. Older belt systems must be adjusted manually.

Belts should be inspected for glazing, rotting, cracking, or swelling. Oil residue on the belt may be an indication of a fluid leak. Also, always check for proper pulley alignment when inspecting or replacing belts. Belt tension must also be checked on both automatic tensioners and on manually adjusted belts. Use a belt deflection gauge or use the deflection method by applying pressure with your finger on the belt midway between the longest span. Compare your reading with manufacturer's specifications. Typical belt deflection is approximately 0.5 inch (12.7mm).

For automatic tensioners, inspect for wear, looseness, and binding. Replace if worn.

Task A.2.5 Diagnose power steering pump noises, vibration, and fluid leakage; determine needed repairs.

The power steering pump supplies the hydraulic pressure needed to operate the steering gear. The most common problem with the pump is seal leaks. Loss of fluid will cause an audible whine or growling noise. Seal leaks will require either pump replacement or an overhaul. If a vibration is felt when turning the steering wheel, check for belt problems, internal pump problems, or loose pump retaining bolts and mounting brackets.

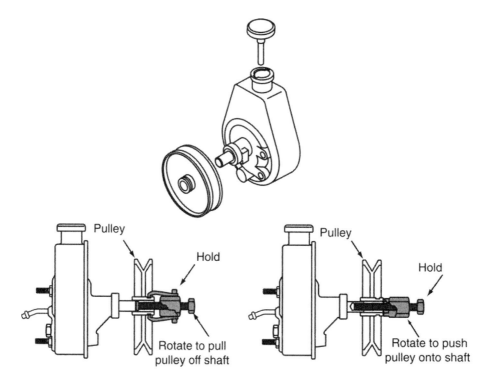

Task A.2.6   Remove and replace power steering pump; inspect pump mounting and attaching brackets; remove and replace power steering pump pulley.

To remove the power steering pump, start by removing the belt and cleaning any dirt around the pump and hose connections. Disconnect the hoses and plug the hoses and pump fittings. Remove the pump mounting bolts and carefully remove the pump from the vehicle. On some vehicles, it will be necessary to remove the pulley before the pump can be removed from the mounting bracket. To remove and install the pulley, always use the proper puller. Never attempt to hammer the pulley off or on. With the pump removed, inspect the mounting brackets.

After installing the pump, fill with fluid, bleed the system, and check for leaks. Also check belt tension and for proper pulley alignment. Road test the vehicle and recheck fluid level and for leaks.

Task A.2.7   Inspect and replace power steering pump seals, gaskets, reservoir, and valves.

Leaking seals and worn parts will require a pump overhaul or replacement. Worn or leaking pumps are usually replaced. However, many times leaking pumps can be resealed. If an overhaul is elected, consult the factory manual first for details and proper procedures.

With the pump removed from the vehicle, drain as much fluid as possible, remove the pulley, and disassemble the pump. Carefully lay out the parts on a clean workbench. Inspect all internal parts, flow control valve, and seals for damage, galling, nicks, and excessive wear.

On pumps with integrated reservoirs, a common location for a leak is from the reservoir seal. Remove the O-ring fitting and mounting bolts. Carefully twist the reservoir and remove the reservoir from the pump housing. Clean the reservoir, mounting area, and seal area. Replace the reservoir seal and reassemble. Lubricate the seal with power steering fluid prior to installation.

The shaft seal is another common cause for leaks. To replace it, first remove the pulley. Pry out the old seal and install the new seal using a suitable seal driver.

When replacing the flow control valve, lubricate the new valve with power steering fluid and install it in the pump housing.

After repairs are made, add new power steering fluid, check for leaks, bleed system, and road test. Check for proper power steering operation. Bleeding procedures are covered in Section A.2.16 of this manual.

Task A.2.8   Perform power steering system pressure and flow tests; determine needed repairs.

Increased power steering effort or lack of power steering assist can be caused by problems in the pump or gear. Checking system pressure can help determine the cause of the problem.

Before starting pressure testing, check fluid level and fill to proper level. Run the engine until normal operating temperature is achieved. Make sure the engine idle is correct and check belt tension. Obtain manufacturer's specifications for the vehicle being tested.

Install a power steering pressure gauge tool on the pressure side between the pump and the gear. Position the shutoff valve toward the power steering gear. Start the engine and record the pressure in the straight-ahead position with the gauge valve open. If the pressure reading is above specification, check for restricted hoses or a damaged steering gear.

Next, turn the steering wheel full left or full right and hold the wheel against the stop for no more than five seconds. Record the maximum pressure attained and check factory specification. Typical pressure readings are generally over 1,000 psi. If the pressure reading is below specification, position the steering wheel in the straight-ahead position and slowly close the shutoff valve. If the pressure rises to the proper specification with the shutoff valve closed, the pump is working correctly. Look for problems in the hoses or gear assembly. If the pressure does not rise with the shutoff valve closed, the problem is in the pump. Do not leave the shutoff valve closed for more than five seconds. If the pressure is okay with the shutoff valve closed, but fails to reach sufficient levels when the steering wheel is turned full left or full right, then the steering gear is faulty.

Task A.2.9   Inspect and replace power steering hoses, fittings, O-rings, coolers, and filters.

Power steering hoses are designed to withstand the extreme high pressure developed by the power steering pump and the high temperatures generated under the hood. Inspect power steering hoses for leaks, cracks, swelling, and physical damage. Always replace a power steering hose with one specifically designed for the vehicle on which you are working. Hoses must be routed and mounted correctly.

Some hose fittings use O-ring seals. Always replace the O-ring when replacing a power steering hose. Tighten the new hose to correct specification and check for leaks.

Many power steering systems use coolers to maintain proper fluid temperature. Inspect coolers for leaks and damage. Filters are often used to trap particles that may damage internal components.

Task A.2.10 Remove and replace steering gear (non-rack and pinion type).

The following are the basic steps to remove and replace a power steering gear.

1. Determine if the steering column must be removed or loosened inside the vehicle. If it must be removed or loosened, proceed as follows. If not, proceed with under-car and under-hood operations beginning with step 2.

    1. Disconnect the battery ground cable or remove the air bag fuse, or fuses, if equipped. If the vehicle has air bags, wait two to five minutes before proceeding.

    2. Disconnect electrical connectors from the steering column under the instrument panel.

    3. Loosen or remove the steering column mounting bolts from the instrument panel bracket.

2. Disconnect the power steering hoses from the steering gearbox and drain the fluid into a suitable container.

3. If necessary for access, remove the power steering pump and any other engine-driven accessories, as required.

4. Disconnect the steering column from the steering gearbox.

5. Remove the Pitman arm from the gearbox sector shaft using a suitable puller.

6. Unbolt the steering gearbox from the chassis and remove it from the vehicle.

7. Reinstallation is the reverse of removal. Properly torque all retainers to specification.

10. After installing the steering gearbox and reconnecting the hoses, fill the pump with fluid and bleed air from the system.

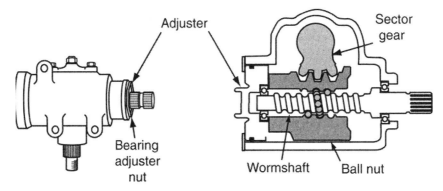

Task A.2.11 Remove and replace rack and pinion steering gear; inspect and replace mounting bushings and brackets.

1. Before removing the rack and pinion gear from the vehicle, point the front wheels in the straight-ahead position.

2. If the vehicle is equipped with an air bag, it is important to lock the steering wheel in place. If the steering wheel rotates while not connected to the gear, damage will occur to the air bag clock spring.

3. Disconnect the steering column coupling or U-joint from the steering gear. Mark the position of the steering column shaft and steering gear to ensure correct alignment during reassembly.

4. Remove the outer tie-rods from the steering knuckle and remove the mounting bolts. The rack and pinion will be either bolted to the frame or to the engine cradle.

5. Remove the rack and pinion gear from the vehicle and carefully inspect for signs of damage and leakage.

6. Check for wear in the rack pinion and inner tie-rods. Inspect the rack bellows boots.

7. Inspect the mounting brackets and bushings for wear. Replace worn or damaged bushings and brackets.

8. When reinstalling the rack, torque all retainers to specification. If the engine cradle was loosened, make sure it is aligned properly during reassembly.

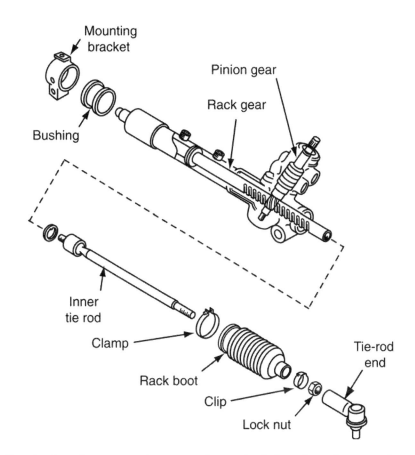

Task A.2.12 Adjust steering gear (non-rack and pinion type) worm bearing preload and sector lash.

Worm bearing preload affects steering effort. Too little bearing preload and the steering will feel loose. Too much preload and the steering will feel tight or stiff. To check worm bearing preload, remove the Pitman arm and the horn pad assembly. With an inch-pound torque wrench, slowly rotate the steering wheel to the left and to the right by using the steering wheel retaining nut. Note the reading while rotating the steering wheel and compare with manufacturer's specification. If an adjustment is needed, loosen the worm shaft bearing locknut. To increase bearing preload, tighten the adjuster nut. To decrease preload, loosen the adjuster nut. After the adjustment is made, turn the steering wheel to the left and to the right. If any roughness or binding is felt, the steering gear may be worn and may have to be replaced. Some systems use shims to adjust preload. Removing shims will increase preload; adding shims will decrease preload.

Sector shaft lash determines how much free play there is in the steering wheel, produced by the steering gear and not by the steering linkage. To check sector lash, remove the Pitman arm and horn pad, as before, and determine the exact center of the steering gear. Rotate the steering wheel to the left one-half turn. Using a inch-pound torque wrench, measure the steering wheel's resistance as it is turned to the right, passing the center point, and continuing until the steering wheel is rotated one full turn. Check your reading against factory specification. If adjustment is required, loosen the sector shaft adjusting lock nut and turn the adjusting screw as required.

Task A.2.13 Inspect and replace steering gear (non-rack and pinion type) seals and gaskets.

Leaking steering gears often means that there is considerable wear, and replacement of the gear is required. Carefully inspect the steering for excessive wear before recommending seals or gasket replacement. Consult the factory manual for procedures and specifications. The steering gear may have to be removed from the vehicle in order to replace some seals.

Thoroughly clean the exterior of the gear before starting the repair. Carefully disassemble the section of the steering gear being repaired and inspect and clean all parts. A scored power steering gear cylinder increases steering effort, but it does not contribute to road shock on the steering wheel. Replace worn seals and gaskets as needed. Use appropriate tools to install seals, as required. After reassembly, check and perform any necessary adjustments and check for leaks.

Task A.2.14 Adjust rack and pinion steering gear.

Two adjustments may be possible on a rack and pinion assembly:

1. Pinion torque is the force needed to turn the pinion gear along the rack. It is adjusted by turning an adjustment screw or a threaded cover on the rack housing or by adding or removing shims under the rack support cover.

2. Pinion bearing preload is the force that the pinion bearings place on the pinion shaft. Only a few steering assemblies have adjustable pinion bearing preload. When it is adjustable, adjustments are made by adding or removing shims or by turning an adjustment collar at the base of the pinion gear.

Most vehicles will require that the rack and pinion assembly be removed for adjustment. On some vehicles, you may be able to disconnect the steering shaft and the tie-rods and make the adjustments on the car. The steering shaft and tie-rods must be disconnected to remove all steering load from the rack and pinion assembly.

Task A.2.15 Inspect and replace rack and pinion steering gear bellows/boots.

The rack and pinion bellows boots protect the inner ball sockets on most designs and the rack seals. The boots contract and expand with the turning of the wheels. Inspect the bellows boots for wear and for fluid seepage. Leaking fluid is an indication of a defective seal or worn rack and pinion. If a bellows boot is cracked, dirt and moisture may have entered. Inspect the inner tie-rod for wear.

To replace the boots, it will be necessary to remove the outer tie-rod and tie-rod locking nut on some models. Mark the position of the tie-rod in order to maintain proper toe alignment during reinstallation. Remove the bellows boot retaining clamps and slide the boot off. Replace the boots using new retaining clamps. On vehicles where the tie-rod is removed, it is important that the wheel alignment be checked after reassembly.

Task A.2.16 Flush, fill, and bleed power steering system.

Flushing the power steering system is accomplished by disconnecting the return to the pump. Plug the pump return port and fill the power steering reservoir with the recommended fluid. With the front wheels off of the ground and the return hose in a drain pan, start the engine and slowly turn the steering wheel from stop to stop. Flushing the system with two quarts of power steering fluid should be sufficient to remove all contaminants and foreign material. Cleaning solvent should never be used in power steering systems for cleaning or flushing procedures. Power steering pumps need power steering fluid to distribute load forces and to prevent excessive heat buildup.

After a repair has been made to the power steering system, bleeding will be necessary to remove trapped air and to obtain a correct fluid level. Refer to the factory manual for bleeding procedures on the specific vehicle being serviced. A typical bleeding procedure is as follows:

- Fill the system with the correct fluid.

- Allow the engine to run for a few minutes without turning the wheels.

- Shut the engine off and let the vehicle sit for a few minutes.

- Recheck the fluid level and refill if needed.

- Start up the engine again and run it for a few minutes. Turn the wheels full left and full right.

- Recheck the level and add fluid, if needed. Repeat the last two steps until the fluid level no longer drops.

- Inspect the entire system for leaks.

- Road test the vehicle and check for correct power steering operation.

- After the road test, check fluid level and add if necessary.

Task A.2.17 Diagnose, inspect, repair, or replace components of variable-assist steering systems.

In a variable-assist steering system with a steering wheel rotation sensor, the hydraulic boost increases when the steering wheel rotation exceeds a specified limit. Power steering assist also is increased at low speeds and decreased at higher speeds.

The system is designed to provide better feel and control at higher vehicle speeds. The variable steering systems are usually designed to start firming up the steering at speeds over 25 mph (40 km/h) and to reach the maximum firmness between 60 and 80 mph (97 and 129 km/h), depending on design. On most vehicles, the main input for the variable-assist steering systems is the vehicle speed sensor, but some manufacturers also use a steering wheel rotation sensor so the vehicle will revert to full assist during evasive maneuvers. On most vehicles, the system goes to full assist below 25 mph (40 km/h).

Pump pressure is controlled by a variable orifice or pressure control valve. Hydraulic pressure is reduced or gradually reduced as vehicle speed increases. Problems within the system may cause a noticeable change in the amount of steering effort at different speeds. The steering may feel stiffer at low speeds or lighter at high speeds. A faulty speed sensor will prevent the system from operating properly. On computerized systems, the controller will put the system in full assist mode at all speeds and steering maneuvers if a fault is detected.

Task A.3   Steering Linkage (3 Questions)

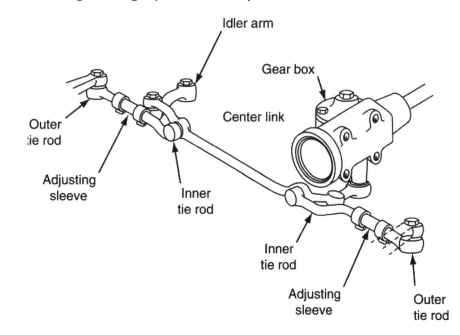

Task A.3.1   Inspect and adjust (where applicable) front and rear steering linkage geometry (including parallelism and vehicle ride height).

The two most common steering linkage designs in use today are the **rack and pinion system** and the **parallelogram design**. The parallelogram is a much more complicated system that uses a conventional steering gearbox.

A variation of the parallelogram design is the cross steer linkage system used on some four-wheel-drive vehicles. The cross steer linkage uses one long tie-rod that is connected to both the left and right steering arm. A drag link connects the left steering arm to the steering gearbox. Some cross steer systems use an adjustable tie-rod drag link, which is used to center the steering wheel.

On all vehicles, a complete periodic inspection of the steering linkage is required due to normal wear and tear of the steering components. All nuts and cotter pins should be in place, and inspected for leaking grease seals. Excessive wear in a steering linkage component will result in inadequate steering performance and can cause tires to wear unevenly.

Also, check the steering linkage for bent or damaged components. A bent component, such as a tie-rod or center link, can affect steering ability and may cause irregular tire wear. Check vehicle ride height when performing a routine linkage inspection. On some vehicles equipped with torsion bars, ride height can be adjusted. On most vehicles, if ride height is incorrect, suspect worn or broken springs. Many vehicles are equipped with rear linkage systems and are susceptible to the same problems as in the front linkage system. Perform a complete inspection of the front and rear linkage systems. On conventional steering linkage (parallelogram), the steering linkage must be parallel to the chassis. Assuming the chassis is level, the simplest method is to measure from both ends of the centerlink to the floor. If the centerlink is unleveled, the idler arm is usually adjustable to correct this problem.

Task A.3.2   Inspect and replace Pitman arm.

The main reason for changing Pitman arms is vehicle crashes. Pitman arms are very well built and the splines do not loosen from road shock. The Pitman arms can be bent or broken in vehicle crashes; but more often sector shafts are bent and broken. If the Pitman arm has a ball and socket at one end, it should be replaced when the ball and socket show any looseness.

To remove the Pitman arm, remove the nut and lock washer first. Some Pitman arms are indexed to the sector shaft. If not, scribe an alignment mark on the Pitman arm before it is removed. This will ensure proper alignment during reinstallation. Using the appropriate puller, remove the Pitman arm. When installing the Pitman arm, make sure it is indexed correctly and torque retaining nuts to specification.

Task A.3.3   Inspect and replace center link (relay rod/drag link/intermediate rod.

A bent or damaged center link can cause incorrect toe alignment, front-wheel shimmy, and possibly tire wear. Inspect the center link for worn joints and for physical damage. To remove the center link, remove the cotter pins and nuts and separate the joints using a separator fork or puller. After installing the center link, properly torque all retaining nuts and replace the cotter pins. If the holes for the cotter pin do not line up after torquing the nuts, continue to tighten the nut until they do. DO NOT loosen the nut to align the hole.

Task A.3.4   Inspect, adjust (where applicable), and replace idler arm(s) and mountings.

The function of the **idler arm** is to hold the right side of the center link level with the left side of the steering linkage. The idler arm is set at the exact same angle, and is the same length as the Pitman arm. The weakest point on the idler arm is the bushing. A worn idler arm bushing will cause excessive movement in the arm, which can cause changes in toe alignment and loose steering. Some idler arms have replaceable bushings; most will have to be replaced if wear is excessive.

To remove an idler arm, disconnect the arm from the steering linkage and unbolt it from the frame. Some idler arms have adjustable slots in the frame. Mark the idler arm prior to removing it to ensure proper positioning when reinstalling. These slots allow adjusting the centerlink to a level position.

Task A.3.5   Inspect, replace, and adjust tie-rods, tie-rod sleeves/adjusters, clamps, and tie-rod ends (sockets/bushings).

The tie-rods connect the center link to the steering knuckle on conventional steering systems. On rack and pinion steering, the tie-rods connect the steering knuckle to the steering gear. Conventional steering systems have inner and outer tie-rods coupled by a sleeve on both sides. The inner tie-rods connect to the center link. On rack and pinion designs, the inner tie-rods connect to the steering gear. Toe alignment adjusters are located on the tie-rod assembly. Toe must be checked after replacement of a tie-rod.

Worn tie-rods can cause front-end shimmy, loose steering, incorrect toe, and may cause tires to wear. Inspect the tie-rod for excessive up and down movement at its ball joint socket. Also check for movement where the tie-rod threads into the adjustment sleeve. Tie-rods with excessive movement will cause changes in toe, affect steering performance and should be replaced.

The tie-rod sleeves must be rotated to adjust front-wheel toe and center the steering wheel.

Replacing the inner tie-rod ends must be done carefully to prevent damage to the pinion teeth on rack and pinion steering. Firmly hold the rack while loosening the socket threads from the threads of the rack.

If the tie rod sleeve clamp is not positioned correctly before tightening, the clamp will not exert enough force to hold the threads together. The constant motion while the vehicle is in operation will wear the threads involved. The two pieces will pull apart, resulting in loss of steering control. On some vehicles, if the tie-rod sleeve clamp is not positioned with the proper orientation, the sleeve bolt could rub against a cross member or a suspension part or wear through a power steering hose.

Task A.3.6   Inspect and replace steering linkage damper(s).

A worn steering damper may transmit road shock to the steering wheel.

Steering linkage dampers (sometimes called steering linkage shock absorbers) are found on vehicles with conventional steering systems. The damper is secured on one end to the vehicle frame and is connected at the other end to the center link.

Steering linkage dampers are designed to help the steering and suspension systems keep road shock and tire vibrations under control. High-speed vibrations are mostly caused by tire and wheel imbalance. Steering linkage dampers work like a shock absorber and are checked the same way one would check for a bad shock.

B. Suspension Systems Diagnosis and Repair (13 Questions)

Task B.1   Front Suspensions (6 Questions)

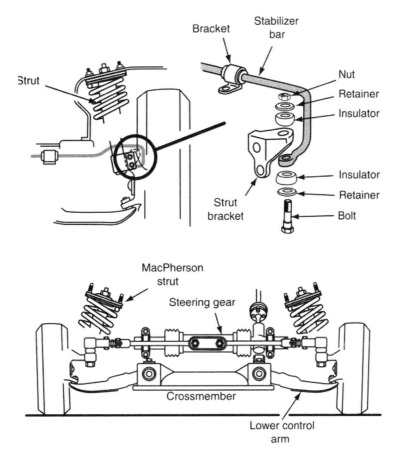

Task B.1.1   Diagnose front suspension system noises, body sway/roll, and ride height concerns; determine needed repairs.

The most common front suspension complaints are noise, steering pull, irregular tire wear, excessive body roll, poor ride quality, and wheel shimmy. These problems are usually the result of worn bushings, worn ball joints, weak springs, faulty shock absorbers, defective tires, or worn steering linkage components. Broken springs or shocks will affect body roll, cause noise, and may affect steering ability. Leaking shocks will not affect vehicle ride height, but may cause poor ride quality.

Road testing the vehicle may reveal abnormal noises, pulling, or steering problems. Pushing down at each corner of the vehicle may uncover noises caused by worn shock bushings, broken springs, or worn suspension components. Wheel bearings and tires should not be overlooked as sources of problems. An out-of-round tire or loose wheel bearing may affect steering handling and ride quality. Also, do not overlook a faulty tire (concentricity) when diagnosing a pull to one side.

Check vehicle ride height on all vehicles and compare it to factory specifications. Incorrect ride height is usually caused by broken or worn springs. Some vehicles are equipped with adjustable torsion bars to correct ride height. Before adjusting the torsion bars, carefully inspect the suspension for worn or bent components. When replacing these bars, note they are designated left and right. Additionally, if reusing an old torsion bar, make index lines prior to removal so the bar can be more easily reinstalled.

Another common source for noise are worn ball joints and control arm bushings. As the vehicle travels over bumps, a squeak or groan can be heard. If the vehicle is equipped with a sway bar, inspect the sway bar links, link bushings, and frame mounts. Broken sway bar links will cause a cracking or banging noise and may cause excessive body roll.

Task B.1.2   Inspect and replace upper and lower control arms, bushings, and shafts.

Control arms allow for the upward and downward movement of the suspension and wheels. The control arms pivot at the ball joints and at the frame. Bushings at the frame allow for this movement. Inspect both lower and upper control arm bushings for wear. Removal of the control arms will be necessary to replace bushings. Some upper control arm designs incorporate a shaft and bushing assembly. This shaft must be carefully inspected and replaced if worn. Control arm bushings are usually replaced with the use of a press. After new bushings are installed, the fasteners are not tightened until the vehicle is on the ground; normal weight of the vehicle is on the suspension and sitting at its normal ride height.

Place safety stands under the lower control arms near the ball joints on most vehicles when replacing the upper control arms because the springs must be partially compressed. When replacing the lower control arms, the safety stands must support the vehicle by the frame so the arms can move down while the springs are removed. The safety stands must be placed in different positions under the vehicle when replacing upper control arms than when replacing lower control arms because of spring location.

Task B.1.3   Inspect and replace rebound and jounce bumpers.

When the wheel hits a bump, the control arms pivot upward causing the spring and shock to compress. Rubber bumpers cushion the blow if the control arms reach their limit of travel. Jounce bumpers make contact if the suspension system is compressed too far. Rebound bumpers compress if the suspension is extended too far. Inspect the rebound or jounce bumpers for wear and cracks. In some cases, the bumper may actually be missing. The bumper may be located on the frame. Or, in the case of MacPherson strut, the bumper may be located on the strut rod under the protective boot. Worn bumpers can be caused by customer driving techniques, or by worn suspension components.

Task B.1.4   Inspect, adjust, and replace strut rods/radius arm (compression/tension), and bushings.

Worn strut rod bushings or a bent strut rod can cause changes in caster and/or toe. Worn bushings may result in the lower control arm moving rearward during braking.

Worn strut rod bushings can cause the vehicle to pull to the direction of the worn bushing every time the brakes are applied. Worn strut rod bushings can cause alignment problems. The adjustments available may not allow the desired specification to be reached if the bushings are worn.

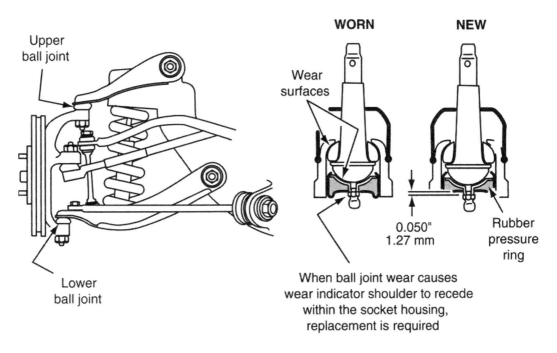

WORN   NEW

Upper ball joint

Lower ball joint

Wear surfaces

0.050"
1.27 mm

Rubber pressure ring

When ball joint wear causes
wear indicator shoulder to recede
within the socket housing,
replacement is required

Task B.1.5   Inspect and replace upper and lower ball joints (with or without wear indicators).

When a coil spring is mounted between the lower control arm and the chassis, position a jack under the lower control arm to unload the ball joints.

Do not place the safety stands under the frame to check for play in the load-carrying ball joint because the front spring tension would make it impossible to measure actual free play. Placing the safety stands under the lower arms on vehicles equipped with MacPherson struts would make ball joint free play impossible to measure because the ball joints would be supporting the weight of the front of the vehicle.

Ball joints carrying the majority of the vehicle weight are known as **load-carrying ball joints.** After correctly positioning the jack stand, check the ball joints for excessive wear. Consult the factory manual for specifications. Some ball joints are designed with a wear indicator at the threaded portion of the grease fitting. If the grease fitting shoulder is receded flush with the outer surface of the ball joint, replace the ball joint.

Worn ball joints can cause alignment problems, tire wear, hard steering, and wheel shimmy. On vehicles with upper and lower ball joints, the load-carrying ball joint usually wears first, but always inspect both. Ball joints must be properly unloaded to check for wear. When the coil spring is located between the lower control arm and the frame, position a jack under the lower control arm. Raise the jack until there is clearance between the floor and the tire. The ball joint is now unloaded. Wear is determined by the amount of axial (up and down) movement of the ball joints. Check the appropriate service manual for vehicle specifications. On MacPherson strut designs, raise the vehicle by the frame until the tire is off the ground and let the lower control hang down.

To replace a ball joint, make sure the control arm is properly supported, either on the control arm or frame, so that the ball joint is not under tension from the spring.

Remove the cotter pin (if equipped) and the ball joint retainer nut. Using a ball joint fork, separate the ball joint. Ball joints can be threaded, riveted, pressed, or bolted in place on the control arm. Use the correct tool, depending on the design. A new threaded ball joint must be torqued to specification. A riveted ball joint will be replaced with a new ball joint supplied with a hardware kit containing bolts, nuts, and washers. Torque the nuts or bolts accordingly. Tighten the ball joint retaining nut to specification and replace the cotter pin. If the hole for the cotter pin does not line up after torquing the nut, continue tightening until it does. Never loosen the nut to align the hole. Grease the ball joint if equipped with a grease fitting.

Task B.1.6   Inspect non-independent front axle assembly for bending, warpage, and misalignment.

Non-independent front axles are primarily used on 4WD trucks. It is a simple and strong design, requiring little maintenance. The non-independent front axle incorporates a solid front axle and is fitted with either leaf or coil springs. The solid axle does not turn with the wheels. The wheels turn by pivoting king pins or ball joints located on the spindle at the ends of the solid axle. Realignment is only necessary if parts are bent or damaged.

A disadvantage of the non-independent front axle is that it provides a rougher ride than independent front suspensions. The up-and-down movement of the front wheels tends to cause a tipping effect and imposes a twisting motion to the frame.

Although durable, a periodic inspection is required on non-independent front suspension. Inspect the axle for warpage, bending, twisting, and physical damage. Damage may not be obvious and looking for abnormal tire wear may sometimes identify a problem with the suspension. Misalignment of the axle will cause tracking problems and tire wear.

Inspect the king pins or ball joints for wear. If equipped with leaf springs, check for proper alignment of the spring to the front axle. If the vehicle design uses coil springs, check for worn or broken springs; check the radius arms and bushings for wear. If the radius arm bushings are badly worn, carefully inspect the radius arm bracket for signs of excessive wear at the point the radius arm passes through the bracket.

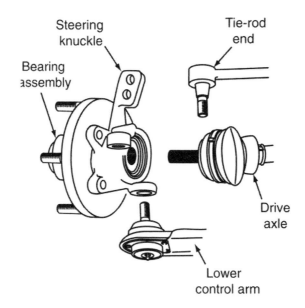

Task B.1.7   Inspect and replace front steering knuckle/spindle assemblies and steering arms.

Excessive tire squeal while cornering may be caused by improper toe-out on turns. This problem can be caused by a bent steering arm. Bent steering arms and steering knuckles or spindles show up in the alignment readings for toe-out on turns and in the steering axis inclination readings. Sometimes a technician can see rust flakes or disturbed metal at the bent section of the part. The parts named must be replaced if they are bent or otherwise damaged.

To remove the steering knuckle, raise the vehicle; remove the wheel and brake components. Support the suspension so that all the tension is removed from the ball joint(s). Disconnect the tie-rod end and separate the ball joints. On some MacPherson strut designs, the strut-to-knuckle bolts are used to adjust camber. Mark the position of the bolts so proper camber can be maintained during reinstallation.

To install the steering knuckle, reverse the procedure, torque all fasteners to specification, and check the front-wheel alignment. Road test the vehicle after installation to ensure proper brake performance and steering operation.

Task B.1.8 Inspect and replace front suspension system coil springs and spring insulators (silencers).

Coil springs are located between the axle (or control arm) and the frame or incorporated on a strut and shock assembly. The spring will compress to support the vehicle at specific ride height. The spring will also compress and rebound in a controlled manner as the vehicle travels over uneven roads. As the springs wear, ride height and ride quality will diminish. Inspect vehicle for signs of leaning to one side, which would indicate a weak or broken spring. Also, check the insulators at the top or bottom of the spring seats. On MacPherson strut design, check the upper strut mount and bearing. Always replace springs in pairs, either front or both rear.

To remove a coil spring, raise the vehicle off the ground and remove the wheel. Disconnect all steering and brake components necessary to gain access to the spring. Compress the spring using the appropriate spring compressor. Support the lower control arm and separate the lower ball joint. It may be necessary to remove the shock absorber. Lower the control arm and remove the spring. With the spring removed, inspect insulators and spring seats for wear. The new spring will have to be compressed to install it. Make sure the new spring is positioned correctly so that it sits correctly on the control arm and/or spring seat. Reinstall all components; drive the vehicle over a bumpy surface to settle the suspension then check the front-end alignment; and road test the vehicle. On MacPherson strut designs, the entire strut and shock assembly is removed as a unit from the vehicle. The spring is compressed and the upper spring mount is removed. Remove the spring and inspect the upper mount, strut bearing, and insulators.

Task B.1.9 Inspect and replace front suspension system leaf spring(s), leaf spring insulators (silencers), shackles, brackets, bushings, and mounts.

Removing and replacing front leaf springs are basic mechanical repair operations. Generally, leaf springs are replaced only when a leaf is broken or when they sag noticeably. Spring are usually replaced in pairs.

A leaf spring is mounted with a rubber bushing and bolt through the eye at one end and by rubber bushings and bolts on shackles at the other end. The shackles at one end of the spring let the spring length change as it flexes. If the spring were mounted directly to the frame at both ends, it would bind and eventually break.

Inspect shackles and bolts for damage and excessive wear. Inspect rubber bushings for wear, deterioration, and damage from grease and oil. Special removal and installation tools often make bushing replacement easier.

Task B.1.10 Inspect, replace, and adjust front suspension system torsion bars and mounts.

A torsion bar performs the same function as a coil or leaf spring. Its function is to support the vehicle weight and allow the wheels to follow the changes in the road surface and also absorb shock. The difference is, unlike a coil spring which compresses, the torsion bar uses a twisting action.

Removing and replacing torsion bars, like springs, are basic repair operations. Torsion bars also are generally replaced only when damaged. Unlike coil and leaf springs, torsion bar stiffness is adjustable on the vehicle; this is what establishes the ride height of the vehicle.

One end of the torsion bar is splined or clamped to a suspension control arm. The other end is secured in a bracket on the chassis. The chassis end of the torsion bar has a short arm and adjusting bolt to set ride height and bar stiffness. Checking the ride height and adjusting it if necessary is a basic part of wheel alignment service. Carmakers' ride height specifications and measurement points vary, so you should check manufacturers' instructions and specifications for this procedure.

Some vehicles are equipped with torsion bars mounted transversely.

If the torsion bars are removed, make sure they are reinstalled on their original side. When replacing torsion bars, check for indicating marks: left and right.

Task B.1.11 Inspect and replace front stabilizer bar (sway bar) bushings, brackets, and links.

Stabilizer bars—also called anti-roll bars or sway bars—minimize body roll, or sway, during cornering. Stabilizer bars do not affect spring stiffness or vehicle spring rate, ride height, or shock absorber action. A stabilizer bar is mounted in brackets with bushings on the car underbody or frame. Links attach each end of the bar to the front or rear control arms or axle housing. During cornering, the bar and its links transfer vehicle loads from the inside to the outside of the suspension. This reduces the tendency of the outside suspension to lift and thus reduces body roll.

The rubber bushings on stabilizer bars and links tend to deteriorate over time and also can be damaged by grease and oil. Worn or damaged bushings should be replaced. Mounting bolts and link bolts may become loose and occasionally break. These should be tightened or replaced as necessary.

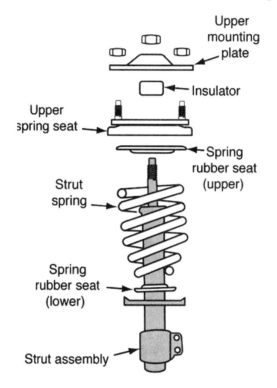

Task B.1.12 Inspect and replace front strut cartridge or assembly.

A new cartridge may be installed in some front struts with the strut installed in the vehicle. Other struts must be removed to allow cartridge installation. Prior to strut removal from the vehicle, remove the upper strut mounting nuts and the strut-to-steering knuckle bolts. Use a spring compressor to compress the spring before the spring is removed from the strut.

MacPherson struts are not only a suspension part, but also serve as a shock absorber and help control vehicle bounce. When replacing just the cartridge and not the outside housing, oil is left in the old housing to help transfer heat.

Task B.1.13 Inspect and replace front strut bearing and mount.

A defective upper strut mount may result in strut chatter while cornering, poor steering wheel return, and improper camber or caster angles on the front suspension.

The caster and camber adjuster plates would make noise if someone had left the bolts loose. The bearings and support plates support the weight of the front of the chassis, the engine, and the transaxle, and have to withstand the weight-shifting forces of braking and the rotating forces of steering the vehicle.

When the steering wheel wants to return to a position other than center, this is known as memory steer. Memory steering occurs when a steering component or bushing binds and prevents the steering gear from smoothly rotating back to center. That is why it is important that all steering components be tightened in their normal resting position. Possible causes for memory steering are: binding upper strut mounts, steering gear, linkage, ball joints, tie-rods, and idler arm. Removing the steering linkage from the steering knuckles can help isolate a binding component.

Sometimes the control valve in the steering gear fails and bypasses fluid into one side or the other of the boost cylinder piston causing the steering to want to turn itself to one side. To check this condition, raise the vehicle off the ground and start the engine. If the wheels want to turn to one side with the engine running, suspect a faulty control valve or steering gear.

Task B.2   Rear Suspensions (5 Questions)

Task B.2.1   Diagnose rear suspension system noises, body sway/roll, and ride height concerns; determine needed repairs.

A squeaking noise in the rear suspension may be caused by suspension bushings, defective struts or shock absorbers, or broken springs or spring insulators. Harsh riding may be caused by reduced rear suspension ride height and defective struts or shock absorbers.

Excessive rear suspension oscillations may be caused by defective struts. Weak coil springs cause harsh riding and reduced ride height. Broken springs or spring insulators cause a rattling noise while driving on irregular road surfaces. Worn-out struts or shock absorbers result in chassis oscillation and harsh riding.

A broken spring leaf will cause the vehicle to lean toward the broken side. Missing insulators will cause creaking and squeaking noises, not rattles, as the suspension moves up and down. Worn shackle bushings or worn shackles will cause rattles when the vehicle is driven over road irregularities at low speeds. Broken center bolts will allow one side of the axle or housing to move forward or rearward. This will change the thrust angle.

Sway bars are not likely to cause vibrations. Coil springs with a high load rating could cause the vehicle to be too high in the rear.

If the rear strut cartridge is weak, the vehicle would bounce more than normal in the rear, but should not hit bottom going over speed bumps at low speeds. If the rear springs are weak, the chassis will hit bottom easily.

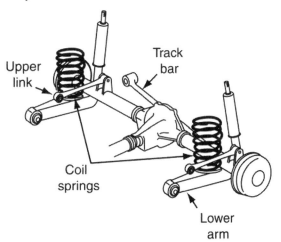

Task B.2.2   Inspect and replace rear suspension system coil springs and spring insulators (silencers).

When replacing rear coil springs, the old spring ends should be matched with the new springs. Matching the spring ends will ensure that the springs are installed correctly. Linear-rate springs or variable-rate springs may be used. Linear-rate springs have equal spacing between the coils and are available as heavy-duty springs for most applications. Variable-rate springs typically have coils spaced closer together at the top with more space between the coils at the bottom of the spring. Variable-rate springs provide automatic load adjustment while maintaining vehicle height. Again these springs are normally replaced in sets.

With the springs removed, inspect the spring insulators and spring seats. Replace worn or cracked insulators. Check for rusted or damaged spring seats. When installing the coil springs, make sure the spring is aligned properly in the spring seat. After the springs are installed, check vehicle ride height and road test the vehicle.

It may be necessary to compress the springs and to disconnect the shock absorber and other components to remove the springs.

Task B.2.3   Inspect and replace rear suspension system lateral links/arms (track bars), control (trailing) arms, stabilizer bars (sway bars), bushings, and mounts.

Different rear suspension designs utilize a variety of components. Generally, rear suspensions use coil springs, leaf springs, or MacPherson struts. When coil springs are used, systems of control arms, links, and/or track bars are used to maintain stability and alignment of the rear axle assembly. The control arms, lateral link, and track bar pivot points are insulated by rubber bushings. Inspect all bushings for wear and check all components for damage, twisting, or bending.

When replacing suspension bushings, it is important that the weight of the vehicle is on the bushings before tightening. Tighten all bushings at their normal standing ride height. Damage to the bushings will occur if they are not tightened in their normal resting position. This procedure applies to track bars and lateral arms also. It may be necessary to use a press or suitable tool to install rear suspension bushings. Only lubricate rubber bushings with an approved lubricant. Never lubricate with oil or grease.

Some rear suspensions have adjustable driveline (pinion) angles. Before installing control arms, check with the factory manual. On some vehicles, an eccentric washer on the control arm adjusts the driveline (pinion) angle.

The rear stabilizer bar (or sway bar) helps to control body roll on turns. The twisting action of the bar counteracts body sway on turns and holds the vehicle closer to a level riding position. Inspect the connecting links, link bushings, and mounting bushings for wear. If the sway bar is bent or damaged, replace it. As with other bushings, tighten only at their normal resting position.

Task B.2.4   Inspect and replace rear suspension system leaf spring(s), leaf spring insulators (silencers), shackles, brackets, bushings, and mounts.

Inspect the leaf springs, bushings, insulators, shackles, and mounts for wear and damage. The most common failure point on a rear leaf suspension is the spring bushings. Check the vehicle ride height to determine the condition of the springs. Leaf springs may sag or break over time. Carefully inspect each leaf in a multiple leaf design for broken leaves, worn insulators, and broken spring retaining clips. On older vehicles, inspect the frame for rot where the spring shackles and mounts bolt up to it. If the springs are not aligned properly with the rear axle or if the rear axle has shifted, check for a broken spring center bolt.

To remove the rear springs, raise the vehicle off the ground, support the rear axle assembly, and remove the wheels. Disconnect the shock absorbers; unbolt the rear shackles and the front mount from the frame. Remove the attaching U-bolts from the axle and remove the springs from the vehicle. Replace all worn bushings and other components, as required. After reinstalling the springs, tighten all bushings with the vehicle weight on the springs and at normal ride height.

Task B.2.5   Inspect and replace rear rebound and jounce bumpers.

The rear rebound and jounce bumpers serve the same purpose as in the front. When the wheel hits a bump, the control arms pivot upward causing the spring and shock to compress. Rubber bumpers cushion the blow, should the control arms reach their limit of travel. Inspect the rebound or jounce bumpers for wear and cracks. In some cases, the bumper may actually be missing.

Task B.2.6   Inspect and replace rear strut cartridge or assembly and upper mount assembly.

Rear struts are serviced similarly to front struts, except that rear struts do not have a steering knuckle. The coil spring on the strut must be compressed to separate it from the strut assembly. Some struts can be serviced by replacing an internal cartridge that contains the shock absorber. Others require replacement of the entire strut.

When reassembling the strut, be sure that the spring is seated securely in its mounting brackets. Inspect the upper mounting location on the car body. Replace any worn or damaged fasteners or other parts. If the body structure is damaged, more extensive repairs will be required.

Task B.2.7   Inspect non-independent rear axle assembly for bending, warpage, and misalignment.

A non-independent rear axle may be checked for bending, warpage, and misalignment by measuring the rear-wheel tracking. This operation may be performed with a track bar or computer wheel aligner with four-wheel capabilities. A track bar measures the position of the rear wheels in relation to the front wheels. A computer wheel aligner displays the thrust angle, which is the difference between the vehicle thrust line and geometric centerline of the vehicle. Rear axle offset may cause steering pull.

On vehicles with rear leaf springs, check the center bolt. If the center bolt is broken, the rear axle assembly may shift, causing misalignment. This would show up while performing a wheel alignment. The thrust angle and rear toe will not be correct. (Thrust angle and toe will be covered in the wheel alignment tasks.)

Task B.2.8   Inspect and replace rear ball joints and tie-rod/toe link assemblies.

Many ball joints have a wear indicator. In these ball joints, the shoulder of the grease fitting must extend a specific distance from the ball joint housing. If this distance is less than specified, replace the ball joint. There is no clearance between the grease fitting shoulder and the ball joint housing.

A worn ball joint may cause improper position of the lower end of the rear knuckle, wheel, and wheel. This action may result in improper rear-wheel camber.

If a rear-wheel tie-rod is longer than specified, the rear-wheel toe-out will be excessive. The length of the tie-rod determines the rear-wheel toe setting.

Rear load-carrying and non-load-carrying ball joints are tested like the front ball joints. Rear tie-rod ends are checked in the same manner as the front tie-rod ends. The rear ball joints and rear tie-rod ends usually last much longer than the front because the rear does not rotate and these components carry much less weight. Some rear ball joints and rear tie-rod ends have to be lubricated.

Task B.2.9   Inspect and replace rear knuckle/spindle assembly.

The steering knuckle or wheel spindle is the mounting point for the wheel and brake assemblies. The wheel rotates on the spindle shaft via a set of bearings. The steering knuckle/spindle is held in place by control arms and/or the suspension strut. To replace a knuckle/spindle, remove the wheel assembly and disconnect the ball joints, control arms, steering linkage, springs, and/or strut assembly from the spindle assembly. After reinstallation, align the wheels.

C. Related Suspension and Steering Service (2 Questions)

Task C.1   Inspect and replace shock absorbers, mounts, and bushings.

The function of the shock absorber is to control and dampen spring oscillations. Since shock absorbers are sealed units, no servicing is required. Inspect the shock bushings and mounts for wear. On most designs, the shock absorber will have to be replaced if the shock bushings are worn. Check for broken mounts and for physical damage to the shock.

When one side of the bumper is pushed downward with considerable force and then released, the bumper should only complete one free upward bounce if the shock absorber or strut is satisfactory. More than one free upward bounce indicates defective shock absorbers, loose shock absorber mountings, or defective struts.

Shocks should be inspected for oil leakage. If oil is dripping from the shock, replace it. The procedure to remove a rear shock absorber is:

1. Lift the vehicle on a hoist and support the suspension on safety stands so the shock absorbers are not fully extended.

2. Disconnect the upper shock mounting nut and grommet.

3. Remove the lower shock mounting nut or bolts.

4. Remove the shock absorber.

When reinstalling, do not over-tighten the mounting hardware to the point that rubber mounting bushings expand past the washer.

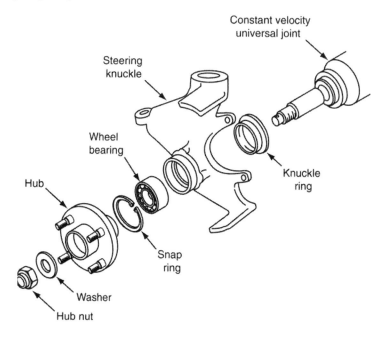

Task C.2   Diagnose and service front and/or rear-wheel bearings.

The wheel bearings perform a major role in effective brake function and steering performance. There are two different types of wheel bearing designs used on modern automobiles: the adjustable tapered roller bearing and the non-adjustable sealed roller or ball bearing.

Worn wheel bearings can cause a growling noise and vibration when the vehicle is driven. Checking for loose or worn wheel bearings should always be part of a complete steering/suspension inspection. A worn or improper wheel bearing adjustment can cause poor brake performance, poor steering, and rapid wheel bearing wear.

Tapered wheel bearings can be disassembled, cleaned, re-greased, and adjusted. Sealed roller or ball bearing wheel bearings cannot be serviced and are replaced as a unit.

Tapered roller bearings are generally used on non-drive axles. The wheel bearings are mounted between a hub and a fixed spindle. To gain access to tapered wheel bearings, remove the wheel, brake rotor or drum, dust cap, cotter pin, spindle nut, and the outer wheel bearing. Remove the hub/rotor or hub/drum assembly. Pry the wheel bearing seal and remove the inner wheel bearing. Thoroughly clean the wheel bearing, hub assembly, spindle shaft, and races. Carefully inspect the wheel bearings, spindle, and races for signs of wear. Discard bearings showing any signs of wear, chipping, galling, or discoloration from overheating. If the race is loose in the hub, the hub is worn and the hub must be replaced. Repack the wheel bearings with high temperature grease. Never repack a wheel bearing without first removing all the old grease. Insert the inner bearing into the hub and lightly lubricate the new wheel seal with grease. Tap the seal in place using a seal driver. Carefully install the hub/rotor or hub/drum assembly onto the spindle. Install the outer wheel bearing, washer, and spindle nut.

If the wheel bearings need to be replaced, it will be necessary to replace the bearing races also. Remove the bearing race from the hub using a bearing race installer. If using a drift punch, tap the races a little at a time, moving the punch around the race to avoid cocking. Use a soft steel drift, or a brass drift, never a hardened punch. When installing the new race, drive the race in until the sound changes, becomes a dull thud, this indicates the race is fully seated.

It is very important to properly adjust tapered wheel bearings. Always check the manufacturer's recommended procedure for the specific vehicle being serviced. There are two widely used methods for adjusting wheel bearings: the torque wrench method and the dial indicator method.

In the torque wrench method, rotate the wheel in the direction of tightening while the spindle nut is tightened to the specified torque. This initial torque setting seats the bearings in the races. The nut is then loosened until it can be rotated by hand. The nut is then re-torqued to a lower specified value. Back the nut, if necessary, to install the cotter pin, and lightly tap on the dust cap.

To adjust the wheel bearings using the dial indicator method, start by tightening the spindle nut while spinning the wheel, to full seat the bearings. Loosen the spindle nut until it can be rotated by hand. Mount a dial indicator so the indicator point makes contact to the machined outside face of the hub. Firmly grasp the sides of the rotor or tire and pull in and out. Adjust the spindle nut until the end play is within manufacturer's specification. Typical end play ranges from .001 inch to .005 inch. Install the cotter pin and dust cap.

Sealed ball and roller bearings are not serviceable and must be replaced when they are worn, defective, or have damaged grease seals. Removing this type of bearing involves pressing the bearing from the hub of the spindle or knuckle assembly. Carefully inspect the bearing, hub, and spindle assembly for wear. Press the new bearing into the spindle/knuckle assembly and torque the axle nut following the manufacturer's procedure. Torquing this nut seats the drive axle in the hub. Note, some bearings are now a self-contained unit and the assembly bolts into the steering knuckle.

Task C.3   ### Diagnose, inspect, adjust, repair, or replace components (including sensors, switches, and actuators) of electronically controlled suspension systems (including primary and supplemental air suspension and ride control systems.

Automatic level control systems are designed to maintain correct vehicle ride height under different load changes. Level control systems use air pressure that is pumped into air shocks or air bag springs in response to different load changes. Air pressure is developed by an electrically operated pump. Some systems use a dryer assembly to absorb moisture from the system. This reduces the chance of corrosion damage to internal components. Some vehicles have a manual shutoff switch to disable the automatic level control system. On these vehicles, the switch must be turned off whenever the vehicle is raised off the ground from the frame with the wheels hanging.

Once the air suspension system has been shut down for an hour, it becomes inactive. If there are leaks, the vehicle ride height will decrease when the vehicle is parked and not in use. It is normal for an air suspension system to drop a little overnight, especially when there is a significant temperature change. If the system is functioning properly, the vehicle will level itself soon after startup. Most air suspension systems, both primary and supplemental, are automatic and have height-level sensors, air control solenoids, relays, an electric air pump, and a module to make the system work. Most modern systems will store fault codes to help with diagnosis. Some systems have a function test that allows each corner of the vehicle to be raised and lowered to verify operation.

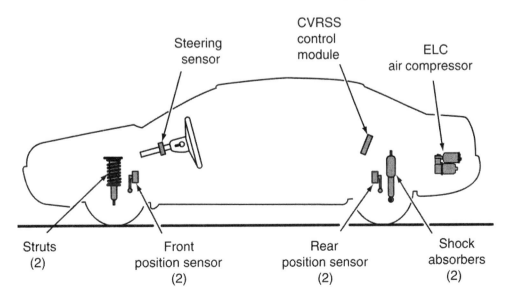

Use a scan tool for diagnosing the electronic suspension. Refer to the scan tool manufacturer's instructions for specific information.

An inspection of an automatic level control system should include: the compressor, height sensors, hoses, hose connections, air shocks (or air struts), electrical connectors, relays, solenoids, wire harness, electrical components, dryer, and pressure regulator. Consult the factory manuals for detailed information on various systems.

The **computerized ride control** system is another automatic suspension design. These suspensions are computer controlled and can adapt to different road conditions and driving situations. Hydraulic pressure is redirected and controlled by the use of actuators within the shock absorber or strut. In this way, the suspension can be altered from a soft ride to a stiffer ride. Some systems will allow the driver to select the type of ride desired for the particular road conditions. For example, a firmer, more controlled setting would be more desirable on a winding road. For highway cruising, a softer mode would be more applicable. On many systems, the computer may override any pre-set modes. If the vehicle is under a hard braking situation, the controller will stiffen up the front shocks to help maintain vehicle control. The same will be true under heavy acceleration. The controller will stiffen the rear shocks to minimize rear-end squatting. When the driver turns hard into a turn, the controller will stiffen the outside shocks and reduce body roll.

Typical computerized ride control systems incorporate a control module, brake sensors, steering sensors, acceleration sensors, mode select switch, actuators, and height sensors. Diagnosing a computerized suspension system varies for different manufacturers. It will be necessary to follow the diagnostic procedures in the appropriate service manual. If the computer recognizes a problem within the system, it will alert the driver by illuminating a warning light on the dashboard. A trouble code may be stored in the memory of the computer, which will aid in the diagnostic process.

Task C.4   Inspect and repair front and/or rear cradle (crossmember/subframe) mountings, bushings, brackets, and bolts.

On unibody vehicles, a subframe is used to help support and locate the drivetrain. On vehicles with a frame and some unibody vehicles, a cross member is used to support the engine and/or transmission. Proper alignment of the drivetrain is critical to the handling of the vehicle and the operation of many systems of the vehicle. As an example, if the cross member or subframe is not secure to the vehicle or if the mounting's bushings are worn, the driver may experience shifting problems due to the misalignment of the shift linkage.

When servicing these units, relieve the weight of the engine and/or transmission before performing any service. This is often done by securing the engine on a hoist.

Task C.5   Diagnose, inspect, adjust, repair, or replace components (including sensors, switches, and actuators) of electronically controlled steering systems; initialize system as required.

The most common electronically controlled steering employs a conventional steering pump and utilizes an electrically operated pressure control assembly on the pump or steering gear. Information from various sensors such as vehicle speed sensor and steering shaft speed sensor are sent to the controller. Steering effort and steering performance is greatly improved. If the controller detects a fault in the system, the steering usually returns to conventional steering assist. A warning light will illuminate should a fault occur; a trouble code may be stored in the computer, depending on the system.

Two new electronically controlled steering system designs are the electro-hydraulic and all-electric steering. These systems are fully computerized and are part of total vehicle management control.

An electro-hydraulic steering system uses a pump driven by an electric motor. The controller will vary pump pressure and flow resulting in varying steering efforts for different driving conditions. The pump can be regulated to run at low speed with low pressure for straight-ahead driving, where full power assist is not needed. During a parking maneuver, pump output will increase, providing increased steering assist where it is needed the most. By controlling the steering assist, steering performance is enhanced and energy is saved.

All electric steering systems use an electric motor that is attached to and operates the steering linkage, steering gear, or steering rack. A variety of motor designs and gear drives are possible, depending on the vehicle. As with the electro-hydraulic steering system, steering effort can be controlled, reducing steering assist while driving straight ahead and increasing steering assist at low speeds.

Typical electronic steering systems incorporate a microprocessor, steering sensor, electric motor, vehicle speed sensor, differential sensor, and other sensors. Inputs from the sensors are sent to the steering control unit and evaluated. Depending on the input from various sensors, the computer will send out the appropriate command to the power unit, which will then control the steering motor. Steering assist will change as the vehicle increases speed, decreases speed, and when the vehicle is turning left or right.

Diagnosing electronically controlled steering starts with a complete visual inspection of the unit, motor, and sensors. Electronic steering systems have the capability of storing diagnostic trouble codes, which will help in analyzing a fault within the system. If a problem does occur, the control unit will shut down the system and employ a fail-safe mode. Steering will revert to manual mode. A warning light will illuminate on the dash panel to alert the driver. Follow the appropriate service manual for diagnostic procedures and for obtaining trouble codes.

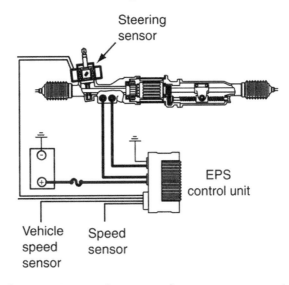

Task C.6   Diagnose, inspect, repair, or replace components of power steering idle speed compensation systems.

Power steering pumps, as in other accessories, put a load on the engine. When the wheels are turned during a parking maneuver and held against the wheel stop, the load increases significantly. Most newer computerized vehicles use a sensor that sends a message to the PCM when power steering pressure is high. The computer will increase the idle to compensate for the load from the power steering pump and prevent stalling.

D. Wheel Alignment Diagnosis, Adjustment, and Repair (12 Questions)

Task D.1   Diagnose vehicle wander, drift, pull, hard steering, bump steer (toe curve), memory steer, torque steer, and steering return concerns; determine needed repairs.

Wheel alignment is the process of measuring and correcting steering and suspension angles. Proper wheel alignment is desired in order to control the vehicle in a safe and predictable manner. Incorrect wheel alignment can cause hard steering, premature tire wear, pulling to one side, wandering, and decreased steering performance. Before a wheel alignment is performed, a complete inspection should be done on the suspension and steering system. Worn steering components or bent suspension parts will prevent the vehicle from being aligned properly. Sagging springs, broken springs, worn wheel bearings, or loose wheel bearings will also affect the wheel alignment.

The first step in the alignment process starts with a road test. Take notice of pulling, wandering, wheel shimmy, and any other steering problems. Does the problem seem to change during braking? Does the vehicle track well in a straight line but feel "loose" when turning? Does it turn better in one direction than the other? These are all items you should be taking in when you are driving the vehicle. Develop a test route and drive it each time to duplicate all kinds of driving conditions and turns. Check and adjust tire pressure while the vehicle is at the correct ride height. Setting the caster, camber, and toe without correcting ride height may not cure tire wear and handling problems. Incorrect ride height in front-wheel drive vehicles may cause vibrations, especially during acceleration. All steering linkage parts, except the idler arm (in some cases), should have zero free play and should be replaced if any looseness is felt or measured during inspection. Incorrect idler arm adjustment can cause bump steer.

In all vehicles, the rear alignment must be correct before any corrections are made to the front. For vehicles with rear-wheel steering, the rear steering system must remain in its centered position while the front adjustments are being checked and adjusted.

Term	Description	Possible Causes
Wander, Drift, and Pull	These terms really describe varying degrees of the same thing. Wander describes a condition in which the vehicle will not travel in a straight line over time. It is usually caused by incorrect settings of caster or toe out. The vehicle will "wander" around the road. Drift is more of a directional wander when a vehicle has a tendency to wander more in one direction. Pull is when a vehicle really leads in a direction as it is moving. Pull may be alignment-related and happens statically (at all times), or may be an intermittent condition that occurs during braking on particular road surfaces. Sometimes switching the tires from side-to-side can help in diagnosing a pull to one side caused by a tire. If the pull goes away or pulls to the other side after switching tires, the problem is with the tires, not the alignment. Since the steering pulls to the side with the least positive caster, excessive positive caster on the left front wheel may cause steering pull to the right. Low tire pressure on the right front will cause the vehicle to pull to the right. Excessive front-wheel toe-in causes feathered tire wear; this problem does not affect steering pull. The steering tends to pull to the side with the most positive camber.	Tire wear Tire cord damage Tire design Inadequate caster setting Incorrect side-to-side caster setting Excessive front or rear toe-out Worn suspension bushings Loose or worn steering components Bent steering/ suspension components Loose wheel bearings
Torque Steer	This condition is caused by uneven application of engine power to the wheels. In front-wheel-drive vehicles, one drive axle is usually shorter than the other. Under acceleration, the vehicle will tend to steer toward the shorter axle. Torque steer can occur in rear-wheel drive and 4WD vehicles when either control arm bushings or spring eye bushings wear and deflect. Torque steer is associated with acceleration or the application of power.	Damaged leaf spring bushings Damaged leaf spring U-bolts Thrust angle problems Incorrect caster/toe settings Worn sway bar bushings (some FWD) Worn radius arm bushings (some FWD) Worn control arm bushings

(Continued)

Term	Description	Possible Causes
Bump Steer (Toe Curve)	This is a dramatic change in the toe settings of the vehicle on one or both wheels when the vehicle encounters a bump or sometimes in the rebound from a bump. The vehicle will demonstrate everything from a small squirm on the road to a dart into the next lane. Bump steer is usually the result of a damaged component in the steering. For example: if a tie-rod is bent on only one side, the toe on that side may change dramatically on a bump. Watch for bump steer on vehicles with altered or sagged ride-height.	Loose rack bushings Loose steering gear Bent steering rod(s) Loose inner tie-rods Sagged springs allowing excessive travel or moving steering out of the designed toe curve.
Memory Steer (hard steering and poor steering return)	This is a term for the steering wheel failing to return to center after a turn. Memory steer can be caused by steering or suspension components that are binding or worn. Many older trucks demonstrated memory steer when their king pins or spindle ball joints were not lubricated regularly. To determine if the problem is in the steering system or the suspension, disconnect the tie-rods from the steering knuckles and manually turn the suspension to isolate the binding component. The failure is almost always in a load-carrying component.	Upper strut bearing failure Seized/dry ball joints or king pins Inadequate positive caster Steering rack or gear adjustment too tight Power steering control valve failure Steering column bearing failure

Memory steering occurs when a steering component or bushing binds and prevents the steering gear from smoothly rotating back to center. That is why it is important that all steering components be tightened in their normal resting position. Possible causes for memory steering are: binding upper strut mounts, steering gear, linkage, ball joints, tie-rods, and idler arm. Removing the steering linkage from the steering knuckles can help in isolating a binding component.

Sometimes the control valve in the steering gear fails and bypasses fluid into one side or the other of the boost cylinder piston causing the steering to want to turn itself to one side. To check this condition, raise the vehicle off the ground and start the engine. If the wheels want to turn to one side with the engine running, suspect a faulty control valve or steering gear.

Task D.2   Measure vehicle ride height; determine needed repairs.

Vehicle ride height is an important specification and must be checked before the alignment is performed. Ride height is usually adjustable on a vehicle with torsion bars.

Ride height can vary significantly on a single model of a light truck with various spring and wheel-and-tire combinations. As ride height varies, so does the front camber angle. Many trucks have different camber specifications for different ride heights. Most truck manufacturers publish tables of varying ride height specifications that should be checked during the alignment operation.

Ride height measurement points vary from one vehicle to another. Some are measured between the lower control arm and the ground; still others have the technician measure between two points of the vehicle chassis. Others are measured between a point on the fender well or under body and ground. Always verify the vehicle manufacturer's measurement points, as well as the specifications.

If ride height is out of limits on a vehicle with coil or leaf springs, the springs or other suspension parts may require replacement. Springs are normally replaced in sets.

Most vehicle manufacturers call for wheel alignment adjustments with the vehicle unloaded and at a specified ride height. Some carmakers, however, specify precise weight loads to be placed in a car during alignment. Trucks are often aligned with specified loads.

Task D.3   Measure front- and rear-wheel camber; determine needed repairs.

Camber is the outward or inward tilt of the wheel as viewed from the top of the tire. The more inward the top of the wheel is from true vertical, the more negative the camber. The more outward the wheel is from true vertical, the more positive the camber. Incorrect camber may cause increased road shock and pulling to one side. Also, incorrect camber can lead to rapid tire wear.

If camber is not adjustable, inspect the struts, suspension, and steering system for bent components. Replace damaged or bent components as needed, and recheck camber angles.

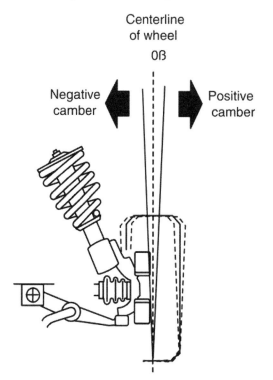

Task D.4   Adjust front- and/or rear-wheel camber on suspension systems with a camber adjustment.

Vehicle manufacturers provide many different ways to adjust front and rear camber:

- Shims may be placed between various suspension components and the frame. (Shims are usually used between control arm pivot shafts and their mounting brackets on the frame.)

- An eccentric cam lobe may be turned to move the control arm pivot point inward or outward on the chassis.

- Sleeves may be adjusted on control arm linkage.

- Strut mounts may be moved.

Vehicle manufacturers publish wheel alignment specifications annually. Most computerized alignment equipment contains an on-board database of specifications and adjustment instructions. Typically a vehicle will have a positive camber setting, and cross camber (side-to-side camber) should not vary more than 0.5 degrees.

Some vehicles have slightly more positive camber on the left front wheel than on the right to minimize vehicle pull caused by the crown of the road. More often, however, road crown compensation is done with slightly different caster angles.

Front-wheel camber and caster are adjusted simultaneously on some vehicles that provide adjustment. Before adjusting camber and caster, jounce the vehicle to relieve any binding or stress on suspension parts and let it settle at its normal ride height. When either angle is adjusted, the other should be checked because changing one will affect the other. The front-wheel toe angle is adjusted after caster and camber adjustments are done. Toe will not change caster and camber settings.

Task D.5   Measure caster; determine needed repairs.

Caster is the tilting of the spindle support centerline from true vertical. Measuring the position of the lower ball joint in relation to the upper ball joint, or the top of the MacPherson strut, determines the caster angle. The spindle support centerline is an imaginary line drawn through the center of the upper ball joint and lower ball joint, or the ball joint and the center of the upper strut mount on a MacPherson strut design. If the lower ball joint is positioned more forward than the upper ball joint, the caster angle is positive. If the lower ball joint is set more rearward than the upper ball joint, the caster angle is negative. On a MacPherson strut design, if the ball joint is positioned more forward than the center of the upper strut mount, the caster angle is positive.

Caster helps improve steering effort, high-speed stability, and steering wheel returnability. Correct caster angles will keep the vehicle traveling in a straight line going forward. On most vehicles, caster is set positive and is usually set at the same degree on both sides or with slightly more on the right side of the vehicle to compensate for the drainage crown built into most roads. Too much positive caster will cause hard (heavy) steering, road shock, and wheel shimmy. Caster set too negative will cause wandering. Caster usually should not vary more than 0.5 degrees from one side to the other. If caster is not within specification and no adjustments are provided, inspect the suspension and steering for bent or damaged components. Also, if the vehicle was involved in a collision, check for frame damage. A vehicle will tend to pull to the side that has the least positive caster. Incorrect caster **will not** cause tire wear.

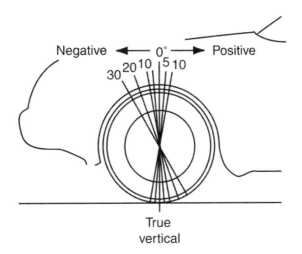

Task D.6   Adjust caster on suspension systems with a caster adjustment.

For vehicles with adjustable caster, consult the service manual for specifications and procedures. Before making caster adjustments, perform a complete steering and suspension inspection. On some vehicles, the caster is adjusted by moving the lower strut rod. By lengthening or shorting the lower strut rod, caster can be brought into specification. On some MacPherson strut vehicles, loosening the upper strut mount and sliding it forward or backward adjusts caster. Vehicles with SLA (short/long arm suspension) often provide caster adjustments. Shimming or sliding the upper control arm will adjust caster. Some upper control arms are designed with eccentric cams that are rotated to adjust caster. Adjusting caster by moving the upper control arm can also incorporate camber adjustment. Some light trucks have an offset upper ball joint bushing that will provide caster adjustment. Bushings come in different sizes that correspond to different degree changes in caster. For example, if you want to change caster 0.5-degrees positive and the bushing in the original upper ball joint bushing is 0 degrees, remove the old bushing and install a 0.5-degree bushing. Make sure the bushing is inserted in the correct position to achieve positive 0.5 degrees. Consult the service manual for procedures on removing and installing the bushing. There are other methods of adjusting caster. Check the service manual for a specific vehicle.

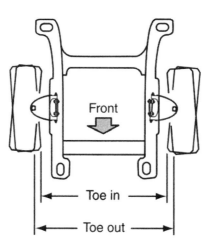

Task D.7   Measure and adjust front-wheel toe.

Setting front toe is performed after camber, caster (front and rear), and rear toe have been checked and /or adjusted. Turn the steering wheel from side-to-side, center it, and then lock it in place using the appropriate tool. If the vehicle is equipped with power steering, have the engine running when centering the steering wheel. Measure front toe and check factory specifications. Front toe is adjusted by turning the toe rods until correct toe is attained. On conventional steering systems, loosen the tie-rod sleeve clamps and turn the sleeve to change toe. For rack and pinion steering, loosen the locknuts and rotate the inner tie-rods. Some vehicles only have one adjusting tie-rod. On these vehicles, the steering wheel is not locked in place. Follow recommended procedures for adjusting toe and centering the steering wheel on these vehicles. On vehicles with adjusting sleeves, make sure the adjusting clamps are re-tightened in the correct position so as not to interfere with other steering linkage or other components when the wheels are turned. After toe has been adjusted, road test the vehicle. The steering wheel should be straight with the wheels in the straight-ahead position.

Task D.8   Center steering wheel.

If front toe is set correctly, the steering wheel should be centered correctly. If the steering wheel is not straight, it may be necessary to recheck toe. Remember to bounce the suspension and to idle vehicles with power steering after toe adjustments to be sure they are correct. Also, make sure the vehicle is tracking correctly. A misaligned rear suspension or twisted frame will affect steering wheel centering. If the steering is not straight, be sure to recheck the rear settings—particularly thrust line.

Task D.9   Measure toe-out on turns (turning radius/angle); determine needed repairs.

The turning radius of the wheel on the inside of the turn is several degrees more than the turning radius on the outside of the turn. When the turning radius is not within specifications, a steering arm may be bent.

Turn the front of the right wheel inward 20 degrees and read the turn table indicator on the left wheel. It should be slightly more than 20 degrees. Perform the same procedure by turning the front of the left wheel inward 20 degrees. Compare these readings and check the service manual for exact specifications. Turning radius angles (toe-out turns) are built into the steering arm design. If the angles are not correct, check the steering arms, steering knuckles, and suspension for bent or damaged components.

Task D.10   Measure SAI/KPI (steering axis inclination/king pin inclination); determine needed repairs.

Steering axis inclination (SAI) is not adjustable, but should be measured and used to analyze handling problems or tire wear problems. Before checking steering angle inclination, completely inspect steering and suspension along with a wheel alignment. SAI is the imaginary line formed by the inward tilt of the upper ball joint or strut mount. The angle is measured in degrees from true vertical. If SAI is incorrect, check for a bent component or frame.

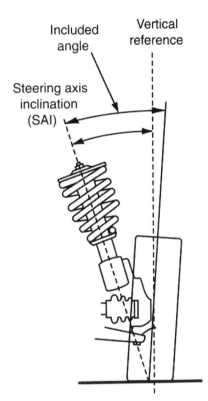

Task D.11   Measure included angle; determine needed repairs.

The **included angle** is determined by adding the camber and steering angle inclination together on any one wheel. A large difference from side-to-side is an indication of a bent suspension component, steering knuckle, or even frame damage. The included angle is not adjustable and is used to determine steering or tire wear problems.

Task D.12   Measure rear-wheel toe; determine needed repairs or adjustments.

Rear toe is usually adjusted by eccentric cams or threaded rods. On some models, rear toe is adjusted by loosening a lock bolt and moving the suspension until the correct toe is attained. Shims installed between the rear axle flange and the wheel bearing can be used to adjust toe when no adjusters are provided. Check with shim manufacturers for specific procedures and application. On models with no adjustments, inspect the rear suspension for misaligned, bent, or damaged components.

On vehicles with **four-wheel steering**, check the appropriate service manual for specific procedures. Usually a special tool is needed to lock the rear steering gear in place while rear toe and rear camber adjustments are made. The tool is removed when the front camber and caster is set, but re-installed to adjust front toe.

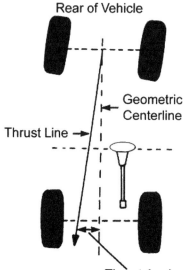

Rear of Vehicle

Geometric
Centerline

Thrust Line →

Thrust Angle

Task D.13   Measure thrust angle; determine needed repairs or adjustments.

Imagine that the vehicle has an exact geometric centerline that runs through the center of it. Now imagine a live axle like you would find in rear-wheel drive vehicles with both rear wheels running exactly parallel to the geometric centerline of the vehicle. This angle of the tires is the thrust angle.

As long as the angle is the same as or very close to the geometric centerline of the vehicle, it will track straight down the road. If the toe is pointed out on one side and in on the other, the vehicle will go in that direction when moving. You have probably seen a vehicle that appears to be going down the road with the rear out to one side or the other of the front wheels. This is often called "dog tracking" and is really a thrust angle problem. Remember that in a rear-wheel drive application, the rear wheels steer the car and the front wheels correct for the rear. In front-wheel drive applications, a vehicle can still "dog track," but it will not normally be perceived by the driver until the rear tires start to wear out. In most situations, one tire will wear more than the other.

Thrust line issues can be caused by only one rear tire being significantly toed in or out. The triangulation of both rear toe readings is used to find the thrust line. Take the point of the triangle, whether it is ahead or behind the vehicle; where it points away from the vehicle centerline is the thrust angle. Causes of thrust angle problems on front-wheel drive or independent rear-end vehicles include: incorrect toe settings; bent rear components like knuckles, control arms, and radius arms; bent or shifted rear suspension cradles; or frame damage. Thrust angle problems on rear-wheel drive live axle vehicles may be caused by: worn spring eye bushings; broken leaf springs or broken leaf spring center bolts; loose or damaged leaf spring U-bolts; worn control arm bushings; bent frame or suspension mounting points; or a bent axle housing.

Task D.14   Measure front wheelbase setback/offset; determine needed repairs or adjustments.

Setback occurs when one front wheel is rearward in relation to the opposite front wheel. Front wheelbase setback is usually caused by collision damage. In other cases, it is designed into the vehicle.

It is possible for caster and camber adjustments to be within specification while the setback is excessive. In that case, the vehicle will pull to the side with the most setback because that side has a shorter wheelbase.

Task D.15   Check front and/or rear cradle (crossmember/subframe) alignment; determine needed repairs or adjustments.

The front cradle may be measured in various locations to verify a bent condition. Some cradles have an alignment hole that must be aligned with a matching hole in the chassis.

The subframe is a critical part of the vehicle suspension and affects the steering. It must be properly centered in its designated location before the mounting bolts are tightened. Vehicle collisions can cause damage to the subframe, which will show up as excessive setback. Subframes must be centered properly with the chassis before the bolts are torqued.

E. Wheel and Tire Diagnosis and Service (5 Questions)

Task E.1   Diagnose tire wear patterns; determine needed repairs.

Feathered tire wear may be caused by improper toe adjustment. Wear on one side of the tire tread usually indicates an improper camber setting. Cupped tread wear may indicate improper wheel balance, worn shocks, or worn suspension components.

Tires wear excessively because the tire tread is not contacting the road surface properly. Alignment adjustments correct most of that problem. Normally, vehicles are aligned with no extra loads in the cargo area or passenger compartment. There are special circumstances, like large drivers or drivers who do not distribute cargo weight properly, that require the vehicle be similarly loaded while the alignment is made.

Task E.2   Inspect tire condition, size, and application (load and speed ratings).

Tires play a vital role in braking performance, handling ability, and ride quality. Defective tires, improper inflation, or installing the wrong tires can have a negative effect on overall vehicle steering and handling.

Different vehicles with different suspensions are designed to be driven with a specific tire construction, tire size, and tire pressure. The load exerted on the tires of a pick-up truck used for hauling heavy cargo is quite different than a small compact sedan. Using the wrong tires on a vehicle or not inflating the tires correctly can cause serious results, tire damage, and may affect steering ability. Always install the correct tire size and adjust pressure to specification. Tire load information labels can generally be found on the door pillar, driver's door, or inside the glove box. The label will have specific information on tire size, inflation pressure, and the maximum vehicle load.

All tires' ratings are determined by a series of numbers located on the tire. The rating system identifies the size, wheel size, type of construction, speed rating, and load rating. A typical tire may have the following designation: **P 225/70 H R 15**. The letter **P** indicates that this tire is designed for passenger vehicles and some light trucks. Other designations are **T** for temporary use, **LT** for light truck, and **C** for large trucks or commercial. The number **225** indicates the width of the tire measured in millimeters. The higher the number, the larger the tire size. The number **70** is the aspect ratio of the tire's height to its width. The higher the aspect ratio number, the taller the tire. The letter **H** designates the speed rating. The rating specifies the maximum speed at which the tire can be driven safely. The rating range is from **B** (31mph) to **Z** (over 149 mph). The letter **R** indicates radial-ply construction. Bias-ply tires are marked **B**. The number **15** is the rim diameter in inches.

Tires are also graded for **temperature resistance**, **traction** and **tread wear**. Temperature resistance is graded on three levels: A, B, and C. "A" provides the greatest resistance to heat; "C" provides the least. Traction is also rated on three levels: A, B, and C. "A" will provide the best traction, especially on wet roads. A "C" rating will give the least amount of traction. Tread wear ratings use a numbering system ranging from 100 to approximately 500. The higher the number, the more mileage can be expected from the tire. A rating of 150 should yield about 50% more mileage than a tire rated at 100.

For safety reasons, a periodic tire inspection should be performed. Check for cuts, bulges, abrasions, stone bruises, and objects embedded in the tire. Most tires have tread wear indicators built into the tread. When the wear indicators appear at two adjacent treads at three or more locations around the tire, replace the tire. If the inner cord or fabric is exposed, the tire must be replaced.

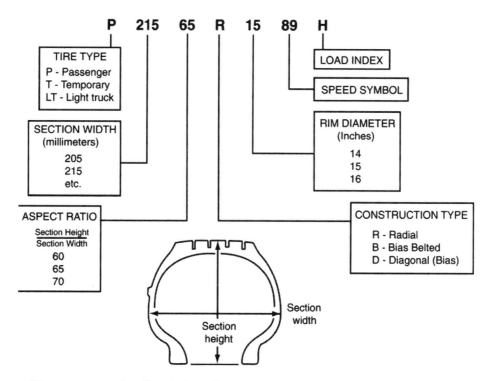

Task E.3   Measure and adjust tire air pressure.

Overinflation causes wear on the center of the tire tread; underinflation causes wear on the edges of the tread. Underinflation may also cause wheel damage. Adjust tire pressure when the tires are cool.

Vans, trucks, and sport utility vehicles commonly require more pressure in the rear tires because of the loads they carry and to minimize sway. Putting more air in the tires on one side of the vehicle is never recommended.

All air pressure specifications on passenger cars and trucks are to be measured while the tire is cold. A cold tire is one that has been driven less than three miles. If the tires have been driven for more than three miles, they must be allowed to cool for two to three hours. Always refer to the vehicle manufacturer's specifications for the tire inflation pressure.

Task E.4   Diagnose wheel/tire vibration, shimmy, and noise concerns; determine needed repairs.

Tire and wheel vibration problems vary, depending upon the particular problem with the tire and/or wheel. Understanding the characteristics of a road test will help in the diagnosis.

Tire-induced vibrations, occur at all speeds. A tire with excessive radial runout (out of round) will cause a low-speed vibration that will seem to go away at highway speeds. A tire with lateral runout will tend to shimmy, especially at low speeds. A bent wheel may also cause a shimmy at all speeds.

Also, inspect the wheels for damage, cracks, and elongated mounting holes. A vibration at 55-65 mph is an indication of a wheel-balance problem.

Problems within the tire, shifting cords, and damaged tires can cause tires and wheels to become unbalanced. Sometimes a thumping noise can be heard from the tires when the tread becomes damaged or chopped. **Wheel tramp** is the term used when the tire hops up and down as it rotates. **Wheel shimmy** refers to the side-to-side shaking that occurs when a tire is not properly balanced. Replace damaged tires, and damaged or bent wheels.

Task E.5   Rotate tires/wheels and torque fasteners according to manufacturers' recommendations.

Most vehicle manufacturers recommend tire rotation at specified intervals to obtain maximum tire life. The exact tire rotation procedure depends on the model year, the type of tire, and whether the vehicle has a conventional or compact spare. For proper tire rotation procedures, refer to the vehicle manufacturer's service manual or owner's manual. There are several different recommended rotation patterns.

Most manufacturers use the modified "X" method which states "cross the non-drive and move the drive tires straight." When rotating tires, inspect the sidewalls of the tires for directional indicators; some tires are designed to rotate in a specific direction. Directional tires are typically rotated front to rear to keep the tires rotating in the correct direction. Some vehicles have directional wheels that are designed to work only on one side or one location on the vehicle. Directional tires offer better handling and response when they roll in the intended direction. They are also constructed with a particular tread design that channels away water more effectively, reducing the chances of hydroplaning.

Tighten all lug nuts to proper torque and sequence. Consult the service manual for specifications. Over tightening the lug nuts can cause damage to the wheel, distort the wheel studs, or warp the hub/bearing assembly.

Task E.6   Measure wheel, tire, axle flange, and hub runout (radial and lateral); determine needed repairs.

Excessive rear chassis waddle may be caused by a shifted steel belt in a tire, or a bent rear hub flange. Tire and wheel runout can be checked by using a runout gauge that follows the tire tread (radial runout), or the gauge can be placed on the sidewall of the tire (lateral runout).

If technicians carefully check runout of all of the parts involved and marks all of the high and low spots, they can correct excessive runout. New parts may not be necessary to correct the problem. Because front-wheel drive cars are lighter and smaller, they transmit noises, vibrations, harshness, and out-of-round conditions more than larger rear-wheel drive cars and trucks.

Task E.7   Diagnose tire pull (lead) problems; determine corrective actions.

Steering pull may be caused by front tires with different types, sizes, inflation pressure, or tread designs, or a front tire with a concentricity defect.

Concentricity is a term used in the tire industry when a tire bead is not centered on the tire body. The bead forms a cone, which causes the vehicle to pull in the direction of the small side of the cone. This condition is more commonly referred to by the technician as tire lead. An out-of-round condition in the rear will cause a vehicle to shake side-to-side. A front out-of-round tire will cause the steering wheel to shake, typically at lower speeds.

Task E.8   Dismount and mount tire on wheel.

Tires are mounted and dismounted from a wheel using a tire changer. A good tire changer allows the tire to be removed without damaging the tire's beads or the edges of the wheel.

When removing or mounting the tire, apply an approved tire lubricant to the bead area. Otherwise excessive strain may be put on the tire bead, resulting in damage to the tire. Clean the sealing area and inspect the wheel for cracks, dents, and burrs. Use a clip-on air chuck while inflating the tire and stand back. Wear safety glasses. Do not exceed 40 psi in an effort to seat the tire bead. If the bead will not seat, deflate the tire and examine the tire for the cause. After the tire is inflated, adjust to recommended pressure and check for leaks.

Task E.9   Balance wheel and tire assembly.

Dynamic wheel balance refers to the balance of a wheel in motion. Cupped tire treads, wheel shimmy, or excessive steering linkage wear may be caused by dynamic wheel imbalance.

A wheel and tire assembly that is statically unbalanced will bounce up and down. A wheel and tire assembly that is dynamically unbalanced will cause the wheel to shake from side-to-side. This is called a shimmy.

To bring a tire in static balance, weight is added to the rim opposite of the heavy side. The amount of weight necessary is determined by the weight of the heavy section of the tire. The added weight will compensate for the heavy part of the tire and bring the tire into correct static balance.

To be in dynamic balance, the tire must also be in static balance. The tire must be spun to check for dynamic imbalance. The wheel-balancing machine will detect any side-to-side imbalance and determine the location and weight to bring the tire in correct dynamic balance. After the weight is added, the tire should be spun again to check for accuracy.

There are many different types of wheel designs used today. Many vehicles are equipped with aluminum wheels. Always use the correct wheel weight designed for the wheel. Using the wrong weight may result in damage to the bead surface and wheel. The wrong design wheel weight may also fail because it may not fit correctly to the contour of the wheel bead area.

Task E.10   Test and diagnose tire pressure monitoring system; determine needed repairs.

The Tire Pressure Monitor System has a sensor in each tire attached to the inside of the rim to monitor the pressures. A light on the dash informs the driver if there is a problem with the tire pressures. When tires are rotated or loss of pressure occurs, the low air pressure light will have to be reset by putting the computer into diagnostic mode according to the manufacturer and the light reset according to manufacturer's procedures. Most cars have a reset mode that allows you to reset the light by turning the key on and letting a small amount of air out of each tire. There are also aftermarket reset tools that are designed to reset the lights the way a manufacturer would.

Sample Test for Practice

Sample Test

Please note the letter and number in parentheses following each question. They match the task in Section 4 that discusses the relevant subject matter. You may want to refer to the overview using the cross-referencing key to help with questions posing problems for you.

1. The steering wheel air bag is being replaced on a vehicle with tilt wheel. All of the following precautions should be observed **EXCEPT:**
   A. always disable the air bag system before servicing an air bag.
   B. always test the air bag by bringing battery power directly to the air bag.
   C. when placing an air bag down on a workbench, face the trim cover up.
   D. discarded air bags must be deployed following manufacturer procedure. (A.1.3)

2. The turning radius on the right front wheel is not within specifications. The cause of this problem could be a:
   A. worn lower right ball joint.
   B. bent steering arm.
   C. worn lower control arm bushing.
   D. worn right stabilizer bushing. (D.9)

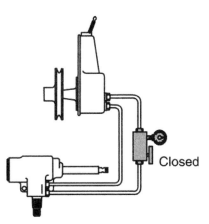

Closed

3. During the power steering pump pressure test shown, the pressure gauge valve should be closed for:
   A. 5 seconds.
   B. 15 seconds.
   C. 20 seconds.
   D. 30 seconds. (A.2.8)

4. Excessive looseness is experienced in a power steering gear (non-rack and pinion type). The cause of this problem could be:
   A. a loose or worn power steering belt.
   B. a loose worm shaft bearing preload adjustment.
   C. a scored steering gear cylinder.
   D. low fluid level in the power steering pump.   (A.2.1)

5. While aligning a vehicle with non-adjustable caster, the technician finds that caster is two degrees less positive on the left wheel and the vehicle is pulling left. Technician A says that the control arms and control arm bushings should be inspected. Technician B says that the tire wear pattern should be inspected for excessive caster wear. Who is right?
   A. A only
   B. B only
   C. Both A and B
   D. Neither A nor B   (D.6)

6. Technician A says front wheel setback is usually caused by worn suspension components. Technician B says a slight front wheel setback causes steering pull. Who is right?
   A. A only
   B. B only
   C. Both A and B
   D. Neither A nor B   (D.14)

7. A recirculation ball type steering gear has a fluid leak from the bottom of the section shaft. Technician A says the seal may need to be replaced. Technician B says the sector shaft busing may need to be replaced. Who is right?
   A. A only
   B. B only
   C. Both A and B
   D. Neither A nor B   (A.2.13)

8. While bleeding air from a power steering system, Technician A says if foaming is present in the reservoir after the bleeding process, the bleeding procedure should be repeated. Technician B says each time the steering wheel is rotated fully right or left, it should be held in this position for two or three seconds. Who is right?
   A. A only
   B. B only
   C. Both A and B
   D. Neither A nor B   (A.2.16)

9. A power steering system has a lack of power steering assist, or hard steering, in both directions. It is determined through testing that the fault is within the non-rack and pinion steering gear. The MOST-Likely cause of the fault is:
   A. a bypassing rotary spool valve.
   B. an over adjusted sector lash.
   C. a bypassing power piston seal.
   D. worm bearing preload that is too loose.   (A.2.1)

10. Technician A says that when adjusting a manual steering gear (non-rack and pinion), sector lash should be set first. Technician B says that worm bearing preload should be adjusted last. Who is right?
    A. A only
    B. B only
    C. Both A and B
    D. Neither A nor B   (A.2.12)

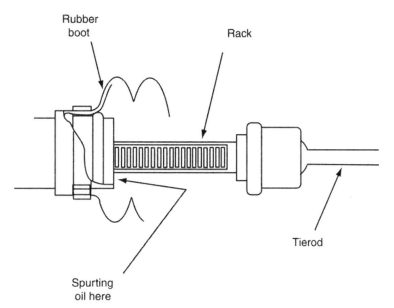

Rubber boot

Rack

Spurting oil here

Tierod

11. A power rack and pinion steering gear has an oil leak at the location shown. The cause of this problem could be a worn:
    A. inner rack seal.
    B. pinion seal.
    C. rack boot.
    D. input shaft seal.   (A.2.2)

12. What could cause air bubbles in the power steering fluid?
    A. Checking the fluid before it reached operating temperature
    B. Overheated fluid
    C. Engine idling too fast
    D. Low fluid level   (A.2.3)

13. A customer complains of a whine noise only when turning. Technician A says the power steering pump may be faulty. Technician B says the left front wheel bearing may be worn. Who is right?
    A. A only
    B. B only
    C. Both and B
    D. Neither A nor B   (A.2.5)

14. To remove the power steering pump, which procedure should be done first?
    A. Remove the Pitman arm.
    B. Remove the belt.
    C. Disconnect the power steering return hose and drain as much fluid as possible; cap all lines.
    D. Remove the power steering pulley.   (A.2.6)

15. On a torsion bar front suspension system, the ride height is below specifications on the right front side of the chassis. The ride height is satisfactory on the left front side of the chassis. Technician A says check for a weak or broken torsion bar on the right side. Technician B says the right front torsion bar anchor bolt may need adjusting. Who is right?
    A. A only
    B. B only
    C. Both A and B
    D. Neither A nor B   (B.1.10)

16. With the steering column mounted in the vehicle and all linkages connected, a steering wheel has excessive free play. Technician A says the flexible coupling may be worn. Technician B says the steering gear mounting bolts may be loose. Who is right?
    A. A only
    B. B only
    C. Both A and B
    D. Neither A nor B   (A.1.2)

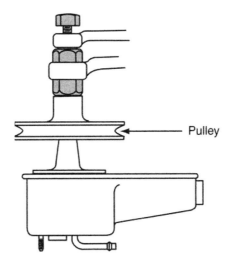

— Pulley

17. The tool shown in the figure is used to:
    A. remove a pressed-on power steering pump pulley.
    B. remove a bolt-on power steering pump pulley.
    C. install a pressed-on power steering pump pulley.
    D. remove the power steering pump pulley retaining nut.   (A.2.6 and A.2.7)

18. Power steering hoses must be replaced if:
    A. fluid is showing at the threaded fitting.
    B. fluid is showing at the molded steel fitting near the end of the hose.
    C. the hose is contacting the body and transmitting noise to the passenger compartment.
    D. a heavy-duty hose is used on a light-duty application.   (A.2.9)

19. The steering does not return to center after a turn on a vehicle equipped with rack and pinion steering gear. Technician A says the steering gear may be misaligned on the chassis. Technician B says the rack pinion preload adjustment may be too tight. Who is right?
    A. A only
    B. B only
    C. Both A and B
    D. Neither A nor B   (A.2.2)

20. A vehicle with power steering pulls to the right. Technician A says that this could be caused by an internal leak in the steering control valve. Technician B says that incorrect wheel alignment could be the cause. Who is right?
    A. A only
    B. B only
    C. Both A and B
    D. Neither A nor B   (A.2.1 and D.1)

21. During a routine steering inspection on a rack and pinion system, the technician finds that a bellows boot is cracked. The technician should:
    A. use a good quality silicone sealer to fill the crack.
    B. ignore it; this is a normal condition.
    C. replace the bellows boot.
    D. inspect and/or replace the inner tie-rod, and replace the bellows boot.   (A.2.15)

22. When replacing a power steering pump belt, the best way to assure proper belt tension is to:
    A. check by hand for 1-inch deflection.
    B. tighten to the specification written on the belt.
    C. use a belt tension gauge.
    D. use a large pry bar to get the belt as tight as possible.   (A.2.4)

23. In a variable-assist power steering system, Technician A says the power steering assist is increased as vehicle speed increases. Technician B says the power steering assist is increased when steering wheel rotation exceeds the specified limit. Who is right?
    A. A only
    B. B only
    C. Both A and B
    D. Neither A nor B   (A.2.17)

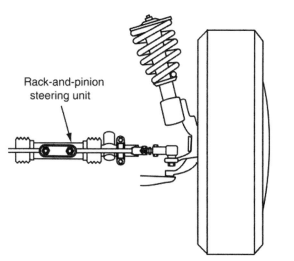

Rack-and-pinion
steering unit

24. An inspection is being performed on the rack and pinion steering system shown in the figure. All of the following should be checked **EXCEPT:**
    A. the ball joints.
    B. the tires.
    C. the Pitman arm.
    D. the tie-rods.   (A.3.1 and A.3.5)

25. Technician A says that excessive steering wheel play can be caused by worn tie-rod ball sockets. Technician B says that steering wheel play can be caused by a worn idler arm. Who is right?
    A. A only
    B. B only
    C. Both A and B
    D. Neither A nor B   (A.3.1 and A.3.5)

26. Technician A says MacPherson strut assemblies are often changed because the vehicle bounces too much on irregular roads. Technician B says on some models, the alignment must be checked after new MacPherson struts are installed. Who is right?
    A. A only
    B. B only
    C. Both A and B
    D. Neither A nor B   (B.1.12)

27. The thrust angle on a front-wheel drive vehicle is more than specified, and the thrust line is positioned to the left of the geometric centerline. This problem could be caused by:
    A. excessive toe-out on the left rear wheel.
    B. excessive toe-out on the right rear wheel.
    C. excessive positive camber on the left rear wheel.
    D. excessive wear in the left rear lower ball joint.   (D.13)

28. During a suspension inspection, the technician discovers a bent center link. Technician A says this problem changes the front wheel toe setting. Technician B says this problem may cause feather-edged front tire wear. Who is right?
    A. A only
    B. B only
    C. Both A and B
    D. Neither A nor B   (A.3.3)

29. Some idler arms are adjustable. Technician A says the adjustment enables settings to allow the vehicle to turn in a tighter circle. Technician B says the adjustment is to give the driver more road feel. Who is right?
    A. A only
    B. B only
    C. Both A and B
    D. Neither A nor B   (A.3.4)

30. During rack and pinion steering gear service, Technician A says the inner tie-rod ends should be replaced if they are worn excessively. Technician B says the rack shaft must be held while loosening the inner tie-rod ends. Who is right?
    A. A only
    B. B only
    C. Both A and B
    D. Neither A nor B   (A.3.5)

31. A four-wheel drive (4WD) vehicle suffers excessive road shock at the steering wheel while driving on irregular road surfaces. Technician A says the power steering gear spool valve may be damaged. Technician B says the steering damper may be worn out. Who is right?
    A. A only
    B. B only
    C. Both A and B
    D. Neither A nor B   (A.3.6)

32. Technician A says power rack and pinion gear assemblies do not have to be adjusted after repair. Technician B says manual rack and pinion gear assemblies must be adjusted after repair. Who is right?
    A. A only
    B. B only
    C. Both A and B
    D. Neither A nor B   (A.2.14)

33. A vehicle has excessive body sway while cornering. All of the following defects could be the cause of the problem **EXCEPT:**
    A. a worn strut rod bushing.
    B. a weak stabilizer bar.
    C. a worn stabilizer bar bushing.
    D. a broken stabilizer link.   (B.1.1 and B.1.2)

34. A customer complains about steering chatter while cornering with a MacPherson strut front suspension system. With the vehicle parked on the shop floor, the technician can feel a binding and releasing action on the left front spring as the steering wheel is turned. Technician A says the upper strut mount may be defective. Technician B says the lower ball joint may have excessive wear. Who is right?
    A. A only
    B. B only
    C. Both A and B
    D. Neither A nor B   (B.1.13)

35. A customer complains of a pull on acceleration and deceleration. Technician A says worn control arm bushings could be the cause. Technician B says a faulty power steering pump could cause this condition. Who is right?
    A. A only
    B. B only
    C. Both A and B
    D. Neither A nor B   (B.1.2)

36. When unloading the ball joints on a front suspension with the coil spring located between the lower control arm and the frame:
    A. a safety stand must be placed under the chassis.
    B. the shock absorber must be disconnected.
    C. the sway bar link bushings must be disconnected.
    D. a safety stand must be placed under the lower control arm.   (B.1.4)

37. A vehicle has excessive tire squeal while cornering. The cause of this problem could be:
    A. a bent steering arm.
    B. excessive negative caster.
    C. worn stabilizer bushings.
    D. worn-out front struts.   (B.1.6 and B.1.7)

38. The tool shown is used to:
    A. compress the coil spring.
    B. compress the coil spring.
    C. measure ball joint movement.
    D. measure lower control arm bushing wear.   (B.1.8)

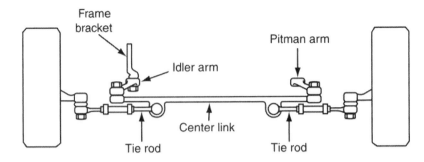

39. In the figure shown, front wheel toe is found to be incorrect. The problem could be all of the following **EXCEPT:**
    A. misadjusted tie-rods.
    B. an incorrectly adjusted steering gear.
    C. a bent center link.
    D. a bent Pitman arm.   (A.3.1 and D.7)

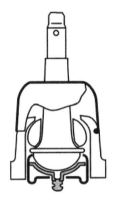

40. While servicing a suspension system as shown, Technician A says the ball joint nut may be loosened to align the cotter pin hole with the nut constellations. Technician B says to torque all ball joint retaining bolts to specification. Who is right?
    A. A only
    B. B only
    C. Both A and B
    D. Neither A nor B   (B.1.5)

41. There is a dull noise and a feeling that something shifted when making a hard turn on a car with rack and pinion steering. Technician A says the rack mounting bushings could be dried out and loose. Technician B says the rack mounting bushings could be oil-soaked. Who is right?
    A. A only
    B. B only
    C. Both A and B
    D. Neither A nor B   (A.2.11)

42. The steering pulls to the right while driving straight ahead on a truck with a long and short arm front suspension and a leaf spring rear suspension. All of these problems could be the cause of the problem **EXCEPT:**
    A. a broken center bolt in the left rear spring.
    B. more positive caster on the left front wheel than right front wheel.
    C. more positive camber on the right front wheel than the left front wheel
    D. excessive toe-in on the front wheels.   (B.1.9)

43. All of the following statements are true of sway bars **EXCEPT:**
    A. the bars are also called stabilizer bars.
    B. if both wheels jounce, the bar pivots in its mounts.
    C. the bar connects the lower control arms to the frame.
    D. the bar transfers the movement of the suspension of one wheel to the suspension of the other wheel.   (B.1.11)

44. The coil spring shown is:
    A. a linear-rate spring.
    B. a variable-rate spring.
    C. a heavy-duty spring.
    D. an adjustable spring.   (B.2.2)

45. A customer complains about steering pull to the left. Technician A says the two front tires may have different tread designs. Technician B says one of the front tires may have a concentricity problem. Who is right?
    A. A only
    B. B only
    C. Both A and B
    D. Neither A nor B   (E.7)

46. The steering pulls to the right while driving straight ahead on a truck with a short/long arm (SLA) front suspension and a leaf spring rear suspension. All of the following defects could be the cause of the problem **EXCEPT:**
    A. a broken center bolt in the left rear spring.
    B. more positive caster on the left front wheel than the right front wheel.
    C. low tire pressure on the right side.
    D. excessive toe-in on the front wheels.   (B.2.4 and D.5)

47. The jounce bumpers are being replaced on the rear of a vehicle. Technician A says the jounce bumper may be mounted on the frame. Technician B says the jounce bumper may be located on the strut. Who is right?
    A. A only
    B. B only
    C. Both A and B
    D. Neither A nor B   (B.2.5)

48. Technician A says the front suspension cradle may be measured at various locations to determine if it is bent. Technician B says on some cradles, an alignment hole in the cradle must be aligned with a hole in the chassis. Who is right?
    A. A only
    B. B only
    C. Both A and B
    D. Neither A nor B   (D.15)

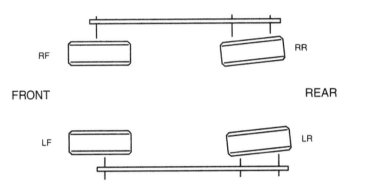

49. A non-independent rear axle is offset as shown in the figure. This problem could result in:
    A. steering wander while driving straight ahead.
    B. the steering wheel being off-center.
    C. poor steering wheel returnability.
    D. steering pull to the left during hard acceleration.   (B.2.7)

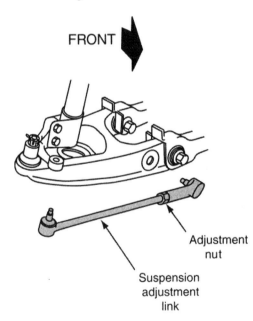

50. The left rear wheel tie-rod shown is longer than specified. This problem could cause:
    A. excessive toe-out.
    B. excessive positive camber.
    C. excessive wear on the outside edge of the tread.
    D. steering pull to the right.   (B.2.8)

51. A vehicle is equipped with a tire pressure monitoring system. Technician A says the sensor must be replaced when the tire is replaced. Technician B says the system must "relearn" after the tires are rotated. Who is right?
    A. A only
    B. B only
    C. Both A and B
    D. Neither A nor B   (E.10)

52. While discussing air shock absorbers, Technician A says that some air shock absorbers can be pressurized with a shop air hose. Technician B says if a shock absorber slowly loses its air pressure, shock absorber replacement may be necessary. Who is right?
    A. A only
    B. B only
    C. Both A and B
    D. Neither A nor B   (C.1)

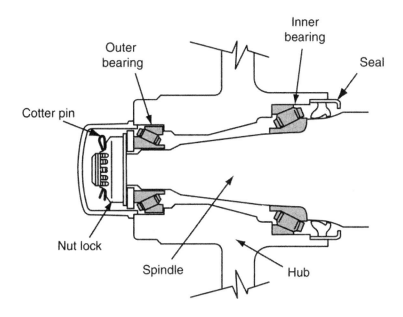

53. When replacing the tapered roller bearing set shown in the figure, Technician A says the specifications for tightening these types of bearings usually leave sufficient free play (approximately .020" - .040") to allow for heat expansion. Technician B says that slight free play is not necessary and the spindle nut should be tightened by hand as much as possible. Who is right?
    A. A only
    B. B only
    C. Both A and B
    D. Neither A nor B   (C.2)

54. To remove the manual steering gear (non-rack and pinion type), which procedure should be done first?
    A. Remove the steering gear from the chassis.
    B. Disconnect the flexible coupling from the worm shaft.
    C. Remove the steering-gear-to-frame mounting bolts.
    D. Disconnect the battery ground cable.   (A.2.11)

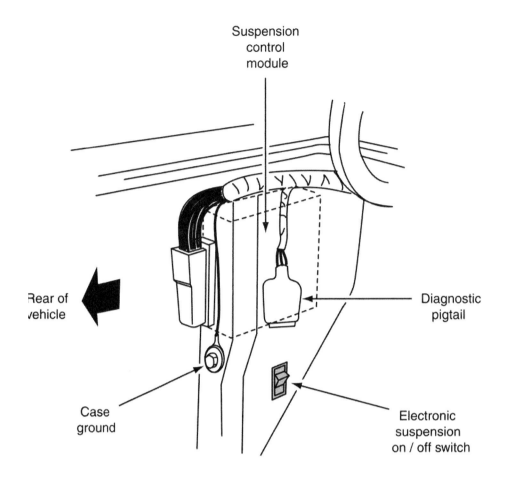

Suspension
control
module

Rear of
vehicle

Case
ground

Diagnostic
pigtail

Electronic
suspension
on / off switch

55. The electronic suspension switch shown must be in the off position and disabled under all
the following conditions **EXCEPT:**
   A. while diagnosing the system with a scan tester.
   B. while jacking the vehicle to change a tire.
   C. while hoisting the vehicle for under-car service.
   D. while towing the vehicle with a tow truck.   (C.3)

56. A vehicle pulls to the right while accelerating. Which of the following could be the cause?
   A. A loose cradle
   B. A loose power steering belt
   C. Both A and B
   D. Neither A nor B   (C.4)

57. A customer is concerned because there is a steering wheel symbol illuminated on the dash at
all times. Technician A says that the lamp means it is time to change the power steering fluid.
Technician B says the lamp means the power steering belt needs to be replaced. Who is right?
   A. A only
   B. B only
   C. C Both A and B
   D. Neither A nor B.   (C.5)

58. A power steering pump with an integral reservoir is leaking fluid between the reservoir and the pump housing. Technician A says the shaft seal may require replacement. Technician B says the fluid return hose may be restricted. Who is right?
    A. A only
    B. B only
    C. Both A and B
    D. Neither A nor B   (A.2.7)

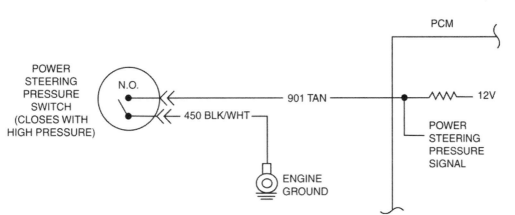

59. Refer to the schematic shown. Technician A says if the power steering pressure switch fails to close with high power steering pressure, the engine may stall. Technician B says if the power steering pressure switch is shorted, there will be 12V on CFT901. Who is right?
    A. A only
    B. B only
    C. Both A and B
    D. Neither A nor B   (C.6)

60. Bump steer is experienced during a road test for steering diagnosis. Technician A says the steering gear may require adjustment. Technician B says the Pitman arm may be bent. Who is right?
    A. A only
    B. B only
    C. Both A and B
    D. Neither A nor B   (D.1)

61. When performing a four-wheel alignment, which procedure should be done first?
    A. Check and/or adjust rear camber and rear toe.
    B. Check and/or adjust front caster, camber, and toe.
    C. Road test, adjust tire pressure, and inspect steering and suspension.
    D. Adjust steering gear preload and lash.   (D.1 and A.3.1)

62. Bump steer is experienced during a road test for steering diagnosis. Technician A says the steering gear may be adjusted too tight. Technician B says the pitman arm may be bent. Who is right?
    A. A only
    B. B only
    C. Both A and B
    D. Neither a nor B   (A.3.2)

63. A customer complains about harsh ride and bottoming of the rear suspension. Technician A says the rear struts may be defective. Technician B says the rear suspension ride height may be less than specified. Who is right?
    A. A only
    B. B only
    C. Both A and B
    D. Neither A nor B   (B.2.1)

64. Technician A says incorrect vehicle ride height can cause tires to wear even though the vehicle has had the caster, camber, and toe set to specifications. Technician B says that incorrect ride height can cause vibrations to come from the drivetrain. Who is right?
    A. A only
    B. B only
    C. Both A and B
    D. Neither A nor B   (D.2 and A.3.1)

65. A vehicle has worn jounce bumpers. Technician A says a low ride height setting could cause this. Technician B says this could be caused by worn shock absorbers. Who is right?
    A. A only
    B. B only
    C. Both and A and B
    D. Neither A nor B   (B.1.3)

66. A vehicle pulls to the left while driving straight ahead. The cause of this problem could be:
    A. inadequate caster on the left front wheel.
    B. excessive positive caster on the left front wheel.
    C. excessive toe-in.
    D. improper toe-out on turns on the right front wheel.   (D.3 and D.5)

67. When one side of the front or rear bumper is pushed downward with considerable weight and then released, the bumper makes two free upward bounces before the vertical chassis movement stops. This action indicates a:
    A. defective shock absorber.
    B. weak coil spring.
    C. broken spring insulator.   (B.2.3 and B.2.6)
    D. worn stabilizer bushing.

68. During alignment of a vehicle with SLA suspension, proper camber cannot be achieved on one of the wheels. Technician A says that the steering axis inclination (SAI) should be checked because a bent spindle could be the cause. Technician B says that the adjustment holes should be elongated to allow for added camber movement. Who is right?
    A. A only
    B. B only
    C. Both A and B
    D. Neither A nor B   (D.3 and D.10)

69. When dismounting and mounting a tire on a wheel, it is recommended to:
    A. use an approved tire bead lubricant.
    B. mount both beads simultaneously.
    C. over inflate to seat stubborn beads.
    D. dismount the tire using only a pry bar.   (E.8)

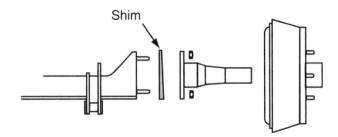

Shim

70. Referring to the figure, Technician A says when replacing the rear spindle, the shim may have to be re-used. Technician B says a four-wheel alignment is recommended, beginning with the rear. Who is right?
    A. A only
    B. B only
    C. Both A and B
    D. Neither A nor B   (D.4 and D.12)

71. All of the following statements about caster adjustment are true **EXCEPT:**
    A. the caster angle is measured with the front wheels straight ahead.
    B. the front wheels are turned a specified number of degrees left and right to read the caster angle.
    C. the brakes must be applied with a brake pedal depressor before reading the caster angle.
    D. the front suspension should be jounced several times before reading the caster angle. (D.5)

72. While performing an alignment on a front-wheel drive vehicle, the technician notices that caster is out of specification and that caster is not adjustable on the vehicle. Which of the following should the technician do first?
    A. Replace both front strut assemblies.
    B. Heat and bend the lower control arms.
    C. Check the cradle (subframe) alignment.
    D. Measure thrust angle.   (D.5 and D.15)

73. Upon road testing a front-wheel drive vehicle, the technician finds that the vehicle must be counter-steered in the opposite direction after every turn. Technician A says that this could be caused by a seized upper strut bearing. Technician B says that this condition is known as memory steer. Who is right?
    A. A only
    B. B only
    C. Both A and B
    D. Neither A nor B   (D.1)

74. Technician A says that as long as the steering wheel spokes are straight, the steering gear is in a center position. Technician B says that to center a steering wheel you must first turn the wheel from stop to stop, counting the number of turns; then turn the wheel back half that amount to obtain a center position. Who is right?
    A. A only
    B. B only
    C. Both A and B
    D. Neither A nor B   (D.8)

75. Technician A says that steering axis inclination (SAI) may need correction if proper camber specifications cannot be achieved during an alignment. Technician B says that correcting SAI may require replacing parts such as the spindle or strut assemblies. Who is right?
    A. A only
    B. B only
    C. Both A and B
    D. Neither A nor B   (D.10)

76. Technician A says included angle is not adjustable and cannot be a problem if it is wrong, so it is of no concern during alignment. Technician B says that included angle is the sum of the camber and SAI measurements. Who is right?
    A. A only
    B. B only
    C. Both A and B
    D. Neither A nor B   (D.11)

77. A front-wheel drive vehicle has excessive toe-out on the left rear wheel. Adjustment shims are positioned between the rear spindles and the spindle mounting surfaces. Technician A says a thicker shim should be installed on the front bolts in the left rear spindle. Technician B says this problem may cause the steering to pull to the right. Who is right?
    A. A only
    B. B only
    C. Both A and B
    D. Neither A nor B   (D.12)

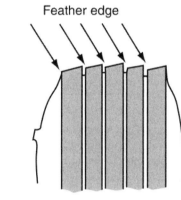

Feather edge

78. The tire tread shown could be caused by:
    A. excessive positive camber.
    B. excessive positive caster.
    C. excessive setback.
    D. improper toe adjustment.   (E.1)

79. A manual rack and pinion steering gear requires excessive steering effort. Technician A says the lower ball joints on the front suspension may be worn. Technician B says the rack bearing adjustment may be too tight. Who is right?
    A. A only
    B. B only
    C. Both A and B
    D. Neither A nor B   (A.2.2)

80. All of the statements about tire inflation are true **EXCEPT:**
    A. overinflation causes excessive wear on the center of the tread.
    B. underinflation causes excessive wear on both edges of the tread.
    C. tire pressure should be adjusted when the tires are hot.
    D. underinflation may cause wheel damage.   (E.2)

81. A vehicle shakes and sways side-to-side at low speeds. Technician A says a defective tire could cause this. Technician B says this problem may be caused by a bent rear wheel. Who is right?
    A. A only
    B. B only
    C. Both A and B
    D. Neither A nor B   (E.6)

82. A vehicle has excessive inside tire wear on the right rear tire. The cause of this problem could be:
    A. a bent rear spindle.
    B. excessive negative caster.
    C. excessive positive camber.
    D. worn-out front struts   (B.2.9)

83. Technician A says tire rotation equalizes wear among the four or five tires rotated. Technician B says different vehicle manufacturers may recommend different rotation patterns. Who is right?
    A. A only
    B. B only
    C. Both A and B
    D. Neither A nor B   (E.5)

6 Additional Test Questions for Practice

Additional Test Questions

Please note the letter and number in parentheses following each question. They match the task in Section 4 that discusses the relevant subject matter. You may want to refer to the overview using the cross-referencing key to help with questions posing problems for you.

1. A vehicle with nonadjustable caster has a severe pull to the left while driving. Technician A says to let some air out of the right front tire and do a road test to see if the pull is still there. Technician B says it could be tire lead and to rotate the two front tires side to side and do a road test to see if pulling to the left is still there. Who is right?
   A. A only
   B. B only
   C. Both A and B
   D. Neither A nor B   (E.7)

2. When replacing directional tires on mating directional wheels, Technician A says the wheel and tire must be mounted to match directional rotation. Technician B says the tire and wheel assembly must be installed on either the left or right side of the vehicle, depending on directional rotation. Who is right?
   A. A only
   B. B only
   C. Both A and B
   D. Neither A nor B   (E.5)

3. Technician A says setback problems can be caused by accident damage to the front cradle. Technician B says most front cradles have a suggested alignment or centering procedure upon installation to ensure that the proper geometric centerline is maintained. Who is right?
   A. A only
   B. B only
   C. Both A and B
   D. Neither A nor B   (D.14 and D.15)

4. To insure correct installation of a new manual or power rack and pinion steering gear, which procedure should be done?
   A. Loosen the outer tie-rod ends and count the number of turns to remove them.
   B. Mark or note the relationship of the steering shaft to the steering gear pinion.
   C. Mark and remove the stabilizer bar mounting bolts.
   D. Mark the position of steering gear mounting bolts.   (A.2.11)

5. Dynamic wheel imbalance on a front wheel may cause all of the following problems **EXCEPT:**
   A. slow steering wheel return to center after turning.
   B. cupped tire tread wear around the tire tread.
   C. front wheel vibration while driving at higher speeds.
   D. excessive wear on shocks/struts.   (E.4 and E.9)

6. Excessive steering effort on a manual steering gear (non-rack and pinion type) may be caused by:
   A. a loose worm bearing preload adjustment.
   B. an overfilled steering gear.
   C. less-than-specified positive caster.
   D. a tight sector lash adjustment.   (A.2.1)

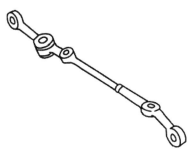

7. The center link shown attaches to the Pitman arm and the idler arm. Technician A says if the center link has a ball-and-socket joint at one end that is loose, the whole center link must be replaced. Technician B says if one of the tapered holes is distorted, the whole center link must be replaced. Who is right?
   A. A only
   B. B only
   C. Both A and B
   D. Neither A nor B   (A.3.3)

8. A vehicle requires excessive steering effort. The power steering belt is tight and the reservoir is filled to the specified level. The cause of this problem could be:
   A. worn lower ball joints.
   B. weak front springs.
   C. a seized idler arm.
   D. a weak stabilizer bar.   (A.2.1 and A.2.2)

9. On a vehicle with non-independent (live axle) rear suspension, the technician notices on the test drive that the steering wheel is biased right. Technician A says the rear axle assembly may be misaligned, causing the thrust line to be off center. Technician B says that one or both of the rear wheels' toe settings could be incorrect causing thrust line to be off center. Who is right?
   A. A only
   B. B only
   C. Both A and B
   D. Neither A nor B   (D.13)

10. The cause of poor returnability on a vehicle equipped with manual rack and pinion steering gear could be:
    A. a loose rack bearing adjustment.
    B. insufficient caster.
    C. loose steering gear mounting bolts.
    D. excessive positive camber on both front wheels.   (A.2.2)

11. The right rear spindle has been replaced on a front-wheel drive vehicle. The technician should:
    A. perform a front-wheel alignment.
    B. perform a four-wheel alignment starting with the rear.
    C. replace all the struts.
    D. bleed the power steering system.   (B.2.7 and D.4)

12. Technician A says if there is oil in the right-hand boot of a rack and pinion gear, both boots need to be changed because the oil can get to both sides through the equalizer tube. Technician B says if the right-hand boot has oil dripping, the rack and pinion should be rebuilt or replaced. Who is right?
    A. A only
    B. B only
    C. Both A and B
    D. Neither A nor B   (A.2.2)

13. During a power steering pump pressure test, the pump will create adequate pressure, but the steering wheel turning effort is excessive. Technician A says the steering gear may be defective. Technician B says the high-pressure hose may be restricted. Who is right?
    A. A only
    B. B only
    C. Both A and B
    D. Neither A nor B   (A.2.8)

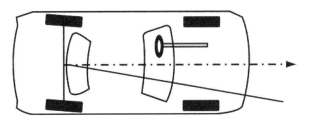

14. During alignment, it is found that the thrust angle is incorrect. Which of the following should the technician do first?
    A. Set front toe to compensate.
    B. Check ride height.
    C. Rotate the tires side-to-side.
    D. Check to see if the rear wheels have a toe adjustment.   (D.13)

15. Technician A says lack of lubricant can seize the king pin, causing hard steering. Technician B says to remove the tie-rod end from the king pin assembly to check king pin pivot movement and ease of travel. Who is right?
    A. A only
    B. B only
    C. Both A and B
    D. Neither A nor B   (A.3.1 and D.1)

16. The power steering rack and pinion gear is being removed on a vehicle equipped with an air bag. The first step the technician would perform is:
    A. disconnecting the outer tie-rods.
    B. disabling the air bag system and locking the steering wheel in place.
    C. disconnecting the power steering lines.
    D. draining the power steering reservoir.   (A.2.11)

17. The power steering pump pulley on a vehicle with a serpentine belt is out of alignment. Technician A says that on many cars, the pulley can be moved into alignment without removing the pump. Technician B says the best way to align the pulleys is to loosen the pump and use shims between the pump bracket and the head/block.
    A. A only
    B. B only
    C. Both A and B
    D. Neither A nor B   (A.2.4)

18. A vehicle has a droning vibration noise only when driven at highway speeds with two people in the back seat. Technician A says it could be a sway bar that is not seated properly. Technician B says the coil springs could be wrong (load rating too high). Who is right?
    A. A only
    B. B only
    C. Both A and B
    D. Neither A nor B   (B.2.1)

19. Shock absorbers are being inspected on a vehicle. Technician A says that a heavy oil film on the housing of the shock absorber indicates the shock is faulty. Technician B says that oil on the outside of the shock does not matter as long as it passes a bounce test. Who is right?
    A. A only
    B. B only
    C. Both A and B
    D. Neither A nor B   (C.1)

20. A vehicle pulls to the right only while braking, and all brake components are in good condition. Technician A says the right front strut rod bushing may be worn. Technician B says there may be excessive negative caster on the right front wheel. Who is right?
    A. A only
    B. B only
    C. Both A and B
    D. Neither A nor B   (B.1.4 and D.1)

21. When removing a non-rack and pinion steering gear on a truck with air bags, Technician A says that while the steering shaft is disconnected, the steering wheel must remain centered. Technician B says that while the steering shaft is disconnected, the air bag clock spring connector should be removed. Who is right?
    A. A only
    B. B only
    C. Both A and B
    D. Neither A nor B   (A.1.3 and A.2.10)

22. A vehicle's ride height is too low. All of the following could be the cause **EXCEPT:**
    A. defective shock absorbers.
    B. a broken coil spring.
    C. a misadjusted torsion bar.   (B.1.1 and B.1.2)
    D. a weak or sagging leaf spring.

23. Stabilizer bars are used on both the front and the rear suspension of many vehicles. Technician A says if inspection reveals the suspension height to be too low on one corner, the stabilizer bar could be the cause. Technician B says if the rear of a vehicle bounces too much while driving, the stabilizer bar could be the cause. Who is right?
   A. A only
   B. B only
   C. Both A and B
   D. Neither A nor B   (B.1.1 and B.1.2)

24. A vehicle with power steering has increased steering effort in both directions. Technician A says that a bent center link could cause this problem. Technician B says that a seized idler arm could cause this problem. Who is right?
   A. A only
   B. B only
   C. Both A and B
   D. Neither A nor B   (A.3.3 and A.3.4)

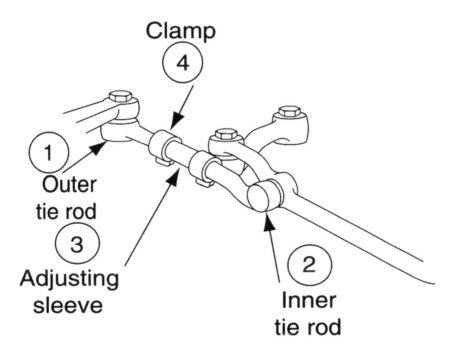

25. Technician A says tie-rod adjusting sleeve clamps, like the ones indicated in the figure, must be positioned a certain way on the sleeve or the sleeve may not grip the threaded end of the tie-rod properly, resulting in loss of steering control. Technician B says on certain vehicles, the tie-rod adjusting sleeve clamps must be properly positioned as shown to prevent interference problems with other suspension parts. Who is right?
   A. A only
   B. B only
   C. Both A and B
   D. Neither A nor B   (A.3.5)

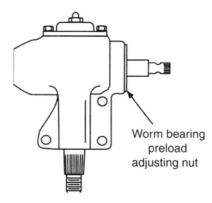

Worm bearing
preload
adjusting nut

26. When adjusting worm bearing preload on a conventional power steering gear, as shown, Technician A says some steering gears require a rotational torque specification to be met. Technician B says that some steering gears require the technician to drain the fluid before performing this adjustment. Who is right?
    A. A only
    B. B only
    C. Both A and B
    D. Neither A nor B   (A.2.15)

27. Technician A says to change the steering linkage damper if it shows signs of leaking or if it has dents in the cylinder housing. Technician B says to change the steering linkage damper whenever there is a vibration in the steering at high speeds. Who is right?
    A. A only
    B. B only
    C. Both A and B
    D. Neither A nor B   (A.3.6)

28. On vehicles equipped with leaf springs, Technician A says the spring eye bushings may wear and can be replaced. Technician B says the interleaf separators can wear or move and may need replacement. Who is right?
    A. A only
    B. B only
    C. Both A and B
    D. Neither A nor B   (B.2.4)

29. Technician A says worn strut rod bushings can cause brake pull problems. Technician B says worn strut rod bushings can cause alignment problems. Who is right?
    A. A only
    B. B only
    C. Both A and B
    D. Neither A nor B   (B.1.4)

30. A customer is concerned about a leak in the air shock system. Technician A says to put 50 psi (345 kPa) air pressure in the system and check for leaks with a soap solution. Technician B says to put 100 psi (690 kPa) air pressure in the system and let the vehicle stand overnight. Who is right?
    A. A only
    B. B only
    C. Both A and B
    D. Neither A nor B   (C.3)

31. Front-wheel shimmy may be caused by:
    A. excessive toe-out.
    B. a broken tire cord.
    C. excessive front wheel setback.
    D. excessive positive camber.   (E.4)

32. During an alignment check of a vehicle with air suspension at all four wheels, the right front height measurement is found to be 0.625 inch (15.9 mm) higher than the left front. Technician A says the height sensor may be disconnected. Technician B says the height sensor adjustment could have slipped. Who is right?
    A. A only
    B. B only
    C. Both A and B
    D. Neither A nor B   (C.3)

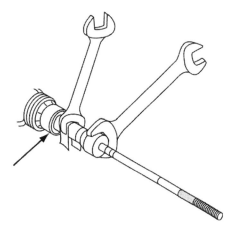

33. In the figure, the component indicated by the arrow is a:
    A. tie-rod.
    B. outer seal bushing.
    C. rack piston ball servo.
    D. inner tie-rod ball socket.   (A.2.11 and A.3.5)

34. A variable assist steering system does not change the steering assist at different road speed changes. Power steering effort appears normal during parking maneuvers. What component might be at fault?
    A. Power steering pump
    B. Steering gear
    C. Rotation sensor
    D. Speed sensor   (A.2.17)

35. When adjusting a torsion bar, the technician should:
    A. remove, heat, twist, and reinstall the torsion bar.
    B. increase or decrease the tension or twist on the bar.
    C. add shims to increase the bar tension.
    D. replace the torsion bar mounting bracket with a bracket that has the specified amount of offset.   (B.1.10)

36. A loud snapping noise can be heard when turning a vehicle to the left or right. This vehicle is equipped with a manual steering gear. What could be the cause?
    A. A bent center link
    B. A loose power steering belt
    C. Worn outer tie-rods
    D. Loose steering gear-to-frame retaining bolts   (A.2.1)

37. Technician A says the tie-rod sleeves may be rotated to adjust front wheel toe. Technician B says the tie-rod sleeves may be rotated to center the steering wheel. Who is right?
    A. A only
    B. B only
    C. Both A and B
    D. Neither A nor B   (D.7 and D.8)

38. When checking a ball joint for radial and axial movement, Technician A says if the lower ball joint is the loaded joint, a jack should be placed under the lower control arm to relieve the load from the joint. Technician B says if the upper ball joint is the loaded joint, a spacer should be placed between the upper and lower control arms and the jack placed under the frame. Who is right?
    A. A only
    B. B only
    C. Both A and B
    D. Neither A nor B   (B.1.5)

39. All of the following are acceptable methods of adjusting camber **EXCEPT:**
    A. adding shims to the upper control arms.
    B. heating and bending the lower control arm.
    C. installing cams at the strut-to-knuckle mount.
    D. installing upper ball joint eccentric spacers.   (D.4)

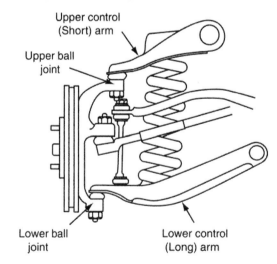

40. When removing a lower control arm on a short arm/long arm suspension system, as shown, the technician should:
    A. compress the spring to remove all spring load from the control arm.
    B. disconnect the ball joint first.
    C. lubricate the control arm bushing with transmission oil for easy removal.
    D. disconnect the outer tie-rod from the steering knuckle.   (B.1.8)

41. When removing a coil spring, Technician A says that before the control arm is disconnected from the steering knuckle, the spring should be compressed with the proper compressor. Technician B says that the shock absorber should be removed before removing the spring. Who is right?
   A. A only
   B. B only
   C. Both A and B
   D. Neither A nor B   (B.1.8)

42. On a rear suspension with two longitudinally mounted leaf springs, the left rear spring is sagged and the left rear chassis ride height is less than specified. This problem could result in:
   A. a change in the front and rear alignment angles.
   B. excessive left rear tire tread wear.
   C. excessive steering wheel free play.
   D. excessive wear to the front steering linkage.   (B.2.4)

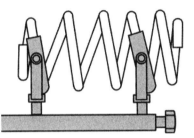

43. While using a spring compressor to remove a coil spring from a strut, as shown, Technician A says the spring should be taped in the areas where the compressor contacts the spring if the compressor does not have protective inserts. Technician B says all of the spring tension must be removed from the upper strut mount before loosening the strut rod nut. Who is right?
   A. A only
   B. B only
   C. Both A and B
   D. Neither A nor B   (B.1.8 and B.2.2)

44. While diagnosing a squeak in a suspension system, it is determined that one of the sway bar-to-frame bushings is making a noise. Technician A says that the sway bar bushing could be replaced to correct the problem. Technician B says that a good quality silicone lubricant could be used on the bushing. Who is right?
   A. A only
   B. B only
   C. Both A and B
   D. Neither A nor B   (B.1.11)

45. Technician A says rear camber and toe should be set to specification, making sure the thrust line setting is at zero before setting front caster, camber, and toe adjustments. Technician B says because the front alignment settings are more important to tire wear, the caster, camber, and toe adjustments on the front wheels should be done before doing the rear wheel adjustments. Who is right?
   A. A only
   B. B only
   C. Both A and B
   D. Neither A nor B   (D.3, D.4, D.6 and D.7)

46. When installing variable-rate coil springs on a rear suspension, all of the following should be done **EXCEPT:**
    A. spring insulators should be placed on the top and bottom of the springs.
    B. the closed coils (coils closest together) should be at the bottom.
    C. closed coils (coils closest together) should be at the top.
    D. springs should be placed so that the end of the spring coils rest in the upper and lower mounts about one-half inch from their stops.   (B.2.2)

47. A vehicle that has been aligned still pulls to the left. Which of the following should the technician do first?
    A. Switch the front tires from side-to-side to determine if the pull switches direction.**Answer A is correct. This is the correct diagnostic procedure.**
    B. Realign the vehicle.
    C. Put maximum recommended air pressure in all the tires.
    D. Retighten all of the wheel nuts to diminish wheel-to-hub runout.   (D.1 and E.7)

48. Specification for total toe on a vehicle calls for 1/8 inch toe-in. Technician A says that while using an alignment machine, each wheel should be adjusted to 1/16 inch toe-in. Technician B says that 1/8 inch toe-in should be set on each wheel. Who is right?
    A. A only
    B. B only
    C. Both A and B
    D. Neither A nor B   (D.7)

49. The steering on a car with a manual rack and pinion steering gear suddenly veers in one direction when one or both front wheels hit a bump. The cause of this problem could be:
    A. a loose steering gear mounting bushing.
    B. excessive positive caster on both front wheels.
    C. worn upper strut mounts.
    D. worn-out front struts.   (D.1)

50. On a vehicle with rack and pinion steering, there is a lack of power steering assist when turning to the left. Right-turn assist is normal. Technician A says that the rack piston seal could be defective. Technician B says that the rotary spool valve seals could be defective. Who is right?
    A. A only
    B. B only
    C. Both A and B
    D. Neither A nor B   (A.2.2)

51. A customer complains about high steering effort and rapid steering wheel return after turning a corner. Technician A says both front wheels may have an excessive positive camber setting. Technician B says the rear suspension ride height may be below spec. Who is right?
    A. A only
    B. B only
    C. Both A and B
    D. Neither A nor B   (D.1 and D.3)

52. A vehicle with rack and pinion steering sometimes darts toward the side of the road after hitting a bump. Technician A says the caster may be too close to zero degrees. Technician B says the rack mounting bushings may be worn and loose. Who is right?
    A. A only
    B. B only
    C. Both A and B
    D. Neither A nor B   (D.1)

53. A vehicle has a noticeable vibration at 60-65 mph. The vibration is felt through the steering wheel and occurs in all gear ranges. Which of the following should the technician do?
    A. Align the front wheels.
    B. Check shock absorber operation.
    C. Balance the wheels.
    D. Balance the drive shaft.   (E.4)

54. Excessive effort is needed to turn the steering wheel on a vehicle with tilt wheel and rack and pinion steering. All of the following are likely causes **EXCEPT:**
    A. worn steering column U-joint.
    B. loose power steering belt.
    C. worn upper strut mount bearings.
    D. worn rack and pinion mount bushings.   (A.1.2)

55. While aligning the front wheels of a vehicle, the technician reads the specification for camber as one-degree positive camber plus-or-minus two degrees. Technician A says that the operating range for camber on this vehicle is from negative one degree to positive three degrees camber. Technician B says that camber may be adjusted anywhere within this range as long as it is equal from side-to-side. Who is right?
    A. A only
    B. B only
    C. Both A and B
    D. Neither A nor B   (D.3 and D.4)

56. A vehicle with a front suspension cradle and strut-type suspension has a two-degree side-to-side camber difference. Which of these is MOST Likely the cause?
    A. Incorrectly positioned cradle
    B. Worn steering rack bushings
    C. Loose wheel bearings
    D. Sagged front springs   (D.11)

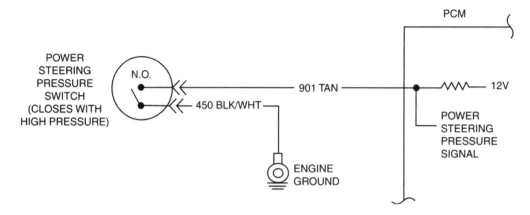

57. Refer to the figure shown. Using a scan tool, the serial data shows the power steering pressure reads high with the wheels straight ahead, at normal idle. The MOST-Likely cause is:
    A. a shorted power steering pressure switch.
    B. an open power steering pressure switch.
    C. an open in CFT 901.
    D. an open in the ground CFT 450.   (C.6)

58. A vehicle pulls right even though all of the alignment adjustments are correct and there is no frame or body damage. Technician A says the front tires could have excessive concentricity. Technician B says one of the tire and wheel assemblies must be severely out-of-round. Who is right?
    A. A only
    B. B only
    C. Both A and B
    D. Neither A nor B   (E.7)

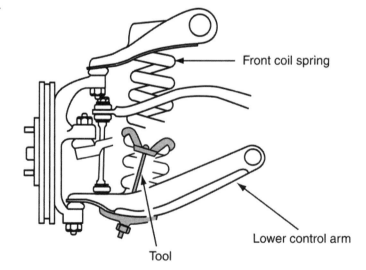

59. All of the following statements about tire inflation are true **EXCEPT:**
    A. overinflation causes excessive wear on the center of the tread.
    B. underinflation causes excessive wear on both edges of the tread.
    C. tire pressure should be adjusted when the tires are hot.
    D. underinflation may cause tire and/or wheel damage.   (E.3)

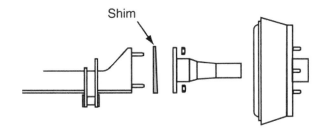

Shim

60. In the figure, the shim installed between the rear spindle and housing is for:
    A. caster adjustment.
    B. turning radius correction.
    C. antilock brake (ABS) sensor adjustment.
    D. toe adjustment.   (D.3 and D.12)

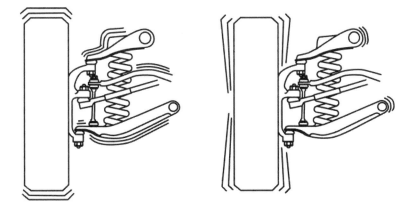

Radial runout   Lateral runout

61. Technician A says that excessive tire radial runout could cause vibration. Technician B says that excessive tire lateral runout could cause wheel shimmy. Who is right?
    A. A only
    B. B only
    C. Both A and B
    D. Neither A nor B   (E.4)

62. When performing an alignment, the technician finds that the included angle is not right. Which of the following should the technician do first?
    A. Check ride height.
    B. Check camber.
    C. Check thrust angle.
    D. Replace the rear springs.   (D.11)

63. Camber is being adjusted on the front wheels of MacPherson strut design suspension. Technician A says that to adjust camber more positive, the top of the wheel must move outward, away from the vehicle. Technician B says to move camber more negative, the lower control arm must move forward. Who is right?
    A. A only
    B. B only
    C. Both A and B
    D. Neither A nor B   (D.4)

64. A vehicle has an excessive amount of tire wear on the outside tread of the right front tire. Technician A says the right front wheel has excessive negative caster. Technician B says that the vehicle has an excessive amount of toe-out. Who is right?
    A. A only
    B. B only
    C. Both A and B
    D. Neither A nor B   (D.4 and E.1)

65. All of the following statements about front wheel toe adjustment are true **EXCEPT:**
    A. the front wheels must be in the straight-ahead position when measuring front wheel toe.
    B. the front wheel toe should be adjusted before the caster angle on the front suspension.
    C. during toe adjustment, the steering wheel must be centered with the front wheels straight ahead.
    D. while adjusting front wheel toe, a tie-rod sleeve rotating tool is used to turn the tie-rod sleeves.   (D.7)

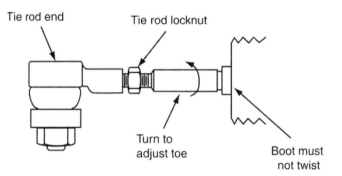

66. After adjusting toe by rotating the inner tie-rod on a rack and pinion system, the technician should:
    A. tighten the outer tie-rod locknuts.
    B. readjust the steering gear.
    C. center the steering wheel and lock it in place.
    D. tighten the idler arm.   (D.7)

67. Front wheel bearings are being serviced on a rear-wheel drive vehicle, and there are metal flakes near the bearing rollers of the outer bearing only. Technician A says the outer bearing and race must be replaced and the other bearings should be cleaned, inspected, and packed with the proper grease. Technician B says all of the bearings on the front wheels should be replaced if one is bad because they have had the same amount of wear. Who is right?
    A. A only
    B. B only
    C. Both A and B
    D. Neither A nor B   (C.2)

68. Technician A says that a variable-assist power steering system is designed to decrease power assist during parking for better control. Technician B says that a variable-assist power steering system is designed to decrease power assist at highway speeds. Who is right?
    A. A only
    B. B only
    C. Both A and B
    D. Neither A nor B   (A.2.17)

69. Technician A says that some steering shaft assemblies use a flexible coupling. Technician B says that worn steering shaft U-joints can cause steering wheel play. Who is right?
    A. A only
    B. B only
    C. Both A and B
    D. Neither A nor B   (D.2 and A.1.2)

70. A vehicle has excessive rear suspension bounce. Technician A says the rear struts may be defective. Technician B says the rear coil springs may be weak. Who is right?
    A. A only
    B. B only
    C. Both A and B
    D. Neither A nor B   (B.2.2 and B.2.6)

71. During inspection of a recirculating-ball manual steering gear, all of the following should be checked for wear **EXCEPT:**
    A. the sector shaft.
    B. the gear housing.
    C. the rack-piston preload.
    D. the Pitman arm.   (A.2.1 and A.2.10)

72. Caster may be adjusted by doing any of the following **EXCEPT:**
    A. adding shims to one or both sides of the control arm.
    B. lengthening or shortening the strut rod.
    C. tightening the torsion bar adjusting bolt.
    D. turning eccentric cams located on the control arm.   (D.6)

Appendices

Answers to the Test Questions for the Sample Test Section 5

1.	B	22.	C	43.	C	64.	C
2.	B	23.	B	44.	B	65.	C
3.	A	24.	C	45.	C	66.	A
4.	B	25.	C	46.	D	67.	A
5.	B	26.	C	47.	C	68.	A
6.	D	27.	A	48.	C	69.	A
7.	C	28.	C	49.	B	70.	C
8.	C	29.	D	50.	A	71.	A
9.	C	30.	C	51.	B	72.	C
10.	D	31.	B	52.	C	73.	C
11.	A	32.	B	53.	D	74.	B
12.	D	33.	A	54.	D	75.	C
13.	A	34.	A	55.	A	76.	B
14.	B	35.	A	56.	A	77.	D
15.	C	36.	D	57.	D	78.	D
16.	C	37.	A	58.	D	79.	B
17.	C	38.	A	59.	A	80.	C
18.	B	39.	B	60.	B	81.	C
19.	B	40.	B	61.	C	82.	A
20.	C	41.	C	62.	B	83.	C
21.	D	42.	D	63.	C		

Explanations to the Answers for the Sample Test Section 5

Question #1
Answer A is wrong. You should always disable the system before working on the steering column or air bag system.
Answer B is correct. Applying battery voltage will deploy the air bag and cause bodily injury to the technician. Always follow the service manual for testing procedures and for the proper use of testing equipment.
Answer C is wrong. Placing the mounting bolts down on the bench is the industry standard method for storing an air bag.
Answer D is wrong. To avoid injury to unsuspecting individuals undeployed air bags must be deployed before disposal.

Question #2
Answer A is wrong. A worn lower ball joint cannot affect the turning radius.
Answer B is correct. This can change toe-out on turns.
Answer C is wrong. A worn lower control arm bushing would only be noticeable in the form of a noise.
Answer D is wrong. Stabilizer bushings have no effect on the turning radius of the vehicle.

Question #3
Answer A is correct. Closing the valve for longer than 5 seconds can cause excessive pressure, damaging the pump.
Answer B is wrong. This could cause internal pump damage.
Answer C is wrong. This could cause internal pump damage.
Answer D is wrong. This could cause internal pump damage.

Question #4
Answer A is wrong. A loose belt does not cause the feel of "looseness" in the steering. It would be hard to steer.
Answer B is correct. Loose preload adjustment can cause excessive steering wheel play.
Answer C is wrong. A scored steering gear cylinder would cause tightness in the steering or binding.
Answer D is wrong. Low fluid level does not cause the feel of "looseness." It would cause more difficult steering effort or loss of power-assist.

Question #5
Answer A is wrong. A bent center link would affect the alignment more than the steering effort.
Answer B is correct. Only Technician B is correct. Any seized part will add resistance to the steering.
Answer C is wrong. Only Technician B is correct.
Answer D is wrong. Only Technician B is incorrect.

Question #6
Answer A is wrong. Setback is either a design feature of the vehicle or caused by collision damage. A bent frame can also cause excessive setback.
Answer B is wrong. Vehicles designed with setback do not suffer steering pull.
Answer C is wrong. Neither Technician is correct. Minor wear in components won't cause set back induced pull.
Answer D is correct. Neither Technician is correct.

Question #7
Answer A is wrong. Technician B is also correct.
Answer B is wrong. Technician A is also correct.
Answer C is correct. Both Technicians are correct. A leak from the sector shaft seal can be caused by a worn seal, however it can also be caused if the sector shaft bushing is worn. If the sector shaft bushing is worn, the sector shaft can move away from the seal when the vehicle is steered and cause the seal to leak. In this instance if the technician replaced only the seal, the root problem has not been corrected.
Answer D is wrong. Both Technicians are correct.

Question #8
Answer A is wrong. Technician B is also correct.
Answer B is wrong. Technician A is also correct.
Answer C is correct. Both Technicians are correct. Foaming indicates air is still in the system. Holding in position for two or three seconds stabilizes pressure, forces air out.
Answer D is wrong. Both Technicians are correct.

Question #9
Answer A is wrong. A bypassing spool valve usually causes hard steering in one direction.
Answer B is wrong. An over adjusted sector shaft will cause binding over center
Answer C is correct. The power piston needs to seal pump pressure to transfer power properly.
Answer D is wrong. Loose worm bearing preload will cause free play in the gear.

Question #10
Answer A is wrong. Sector lash should be adjusted last, since this would remove excessive free play.
Answer B is wrong. Worm bearing preload should be adjusted first. This would cause memory steer if it is too tight.
Answer C is wrong. Neither Technician is correct.
Answer D is correct. Neither Technician is correct.

Question #11
Answer A is correct. The inner rack seal is leaking. When diagnosing a rack and pinion leak, the most common initial cause for premature failure of internal seals and components is dirt and debris entering the inner seals as a result of the outside rack shaft boot being torn or coming loose.
Answer B is wrong. The location shown is nowhere near the pinion shaft.
Answer C is wrong. The rack boot protects the rack and seal from debris. No fluid should be present in the boot.
Answer D is wrong. The input shaft is the same as the pinion shaft.

Question #12
Answer A is wrong. Cold fluid would not be aerated.
Answer B is wrong. The fluid cannot become aerated if the fluid is overheated.
Answer C is wrong. Engine speed cannot cause the power steering fluid to become aerated.
Answer D is correct. Low fluid level can cause the pump to pull in air and cause the fluid to become aerated.

Question #13
Answer A is correct. Only Technician A is correct. A whine noise only while turning can be caused by a faulty power steering pump. When the steering wheel is turned the pump is required to create more pressure, in a worn pump this may cause a whine sound.
Answer B is wrong. Technician B is wrong. Wheel bearing noise will be more pronounced while turning. However it will not be a whine noise, it will be a growling noise.
Answer C is wrong. Only Technician A is correct.
Answer D is wrong. Only Technician A is correct.

Question #14

Answer A is wrong. The Pitman arm is connected to the steering gear and is not removed with the pump. The first thing that needs to be done is to loosen the belt tensioner and remove the belt. Removing the pulley, return hose, and draining the fluid are steps performed after the belt is removed.

Answer B is correct. Removing the belt should be done first to keep the belt from getting fluids on it. As with almost any repair, disconnect the battery for safety. To remove most power steering pumps, refer to the service manual for disassembly. Once the belt is removed, disconnect the pressure and return lines and drain the fluids. On some vehicles, you will need to remove the pulley to access the bolts that secure the pump in place. Some may also have a support bracket on the rear of the pump. Once the attaching bolts are out, the pump can be removed from the vehicle.

Answer C is wrong. This is usually performed after belt removal.

Answer D is wrong. You have to remove the belt before you can remove the pulley. Also, belt alignment should be checked before removal.

Question #15

Answer A is wrong. Technician B is also correct.

Answer B is wrong. Technician A is also correct.

Answer C is correct. Both Technicians are correct. A weak or broken torsion bar may affect ride height on one side. Technician A is correct. A misadjusted torsion bar can affect ride height. Torsion bar suspension uses a long, round piece of spring steel that is splined on one end and notched on the other end to support the body off the frame. This spring steel rod is made specifically for the left or right side of the suspension; this rod cannot be interchanged or it will break. The torsion bar is usually connected to the lower control arm. The other end is attached to an adjusting lever that is attached to the transmission cross member or vehicle frame . This is where the ride height is adjusted.

Answer D is wrong. Both Technicians are correct.

Question #16

Answer A is wrong. Technician B is also correct. The excessive free play could be caused by loose steering gear mounting bolts.

Answer B is wrong. Technician A is also correct. The excessive free play could be caused by a worn flexible coupling.

Answer C is correct. Both Technicians are correct. Carefully inspect mounting bolts and check for proper torque, in addition to the condition of the flexible (U-joints) coupling. There also may be a noise while moving the steering wheel.

Answer D is wrong. Both Technicians are correct.

Question #17

Answer A is wrong. The power steering pump pulley is removed with a puller that fits into the groove on the pulley.

Answer B is wrong. Removing a bolt-on pulley does not require a special tool.

Answer C is correct. This insures the pulley is not cocked.

Answer D is wrong. The pulley shown in the illustration is pressed onto the pump shaft. No retaining bolt or nut is required on this particular design.

Question #18

Answer A is wrong. If fluid is leaking from the thread fitting, tighten the fitting or replace the seal.

Answer B is correct. There is no repair for this type of assembly, except replacement.

Answer C is wrong. If the hose is transmitting noise, move the hose so that it does not touch the body.

Answer D is wrong. A heavy-duty hose can be used on a light application; the hose will just last longer.

Question #19

Answer A is wrong. Misalignment of the rack would cause toe-angle alignment problems, but not returnability problems.

Answer B is correct. Only Technician B is correct. Excessive rack pinion reload could restrict rack movement and cause returnability problems.

Answer C is wrong. Only Technician B is correct.

Answer D is wrong. Only Technician B is correct.

Question #20
Answer A is wrong. Technician B is also correct.
Answer B is wrong. Technician A is also correct.
Answer C is correct. Both Technicians are correct. Some power steering systems have an adjustable power steering valve. If the alignment is correct and there is no tire causing a pull, there may be a problem with an internal leak or adjustment.
Answer D is wrong. Both Technicians are correct.

Question #21
Answer A is wrong. Using sealer is not the proper way to repair a bellows boot.
Answer B is wrong. Ignoring the problem will not fix it.
Answer C is wrong. Dirt and moisture may have entered the inner tie-rod.
Answer D is correct. Since the boot is cracked dirt may have entered wore the inner tie-rod end.

Question #22
Answer A is wrong. Deflection is not as accurate for measuring tension as a belt tension gauge is. belt tension.
Answer B is wrong. Specifications for belt tension are in service manual only.
Answer C is correct. A belt tension gauge is the best way to assure proper belt tension.
Answer D is wrong. This can damage components due to excessive load on bearings.

Question #23
Answer A is wrong. The power steering assist is increased only at low vehicle speeds; at higher speeds the assist is reduced.
Answer B is correct. Only Technician B is correct. The system recognizes steering wheel speed (rate or change) and varies the amount of assist the system will provide, therefore reducing effort.
Answer C is wrong. Only Technician B is correct.
Answer D is wrong. Only Technician B is correct.

Question #24
Answer A is wrong. Ball joints should be inspected within the steering system for wear and looseness.
Answer B is wrong. Tires should be inspected along with the steering system for abnormal wear.
Answer C is correct. A rack and pinion steering system does not have a Pitman arm.
Answer D is wrong. Tie-rods should be inspected as part of the steering system for wear and looseness.

Question #25
Answer A is wrong. Technician B is also correct. Excessive steering play can also be caused by a loose idler arm.
Answer B is wrong. Technician A is also correct. Worn tie rod ball sockets can cause excessive steering play.
Answer C is correct. Both Technicians are correct. Worn steering components can cause excessive steering play.
Answer D is wrong. Both Technicians are correct.

Question #26
Answer A is wrong. Technician B is also correct.
Answer B is wrong. Technician A is also correct.
Answer C is correct. Both Technicians are correct. Struts, like shock absorbers, need to be replaced if they allow too much body bounce or travel. Technician B is also correct. On some models, the strut to spindle attaching bolts may also incorporate adjustable cams, which is used for camber adjustment. Both technicians are correct.
Answer D is wrong. Both Technicians are correct.

Question #27

Answer A is correct. Excessive toe-out on the left rear wheel will position the thrust angle to the left of the geometric centerline. *Thrust angle* on a vehicle is a term used in alignments of the suspension system. If a line were drawn down the center of the vehicle, looking down from the top, it would turn left or right according to the direction that the rear wheels are pointing. This can be checked with an alignment machine. If the thrust angle is out of specification, it would make the steering wheel off center and cause the vehicle to "dogtrack," or look like it's going down the road sideways. The front and back wheels would not follow the same track down the road.

Answer B is wrong. Excessive toe-out on the right rear wheel will position the thrust angle to the right of the geometric centerline.

Answer C is wrong. Positive camber will not place the thrust angle out of allowable specifications.

Answer D is wrong. Wear in a rear ball joint should not place the thrust angle out of allowable specifications.

Question #28

Answer A is wrong. Technician B is also correct. A bent center link can cause the toe angle to change, resulting in feather-edged tire wear.

Answer B is wrong. Technician A is also correct. A bent center link will change the front wheel toe angle, resulting in feather-edged tire wear.

Answer C is correct. Both technicians are correct. A bent center link will change the front wheel toe angle, resulting in feather-edged tire wear.

Answer D is wrong. While inspecting the front of a vehicle's suspension with a center link connecting the two tie rods to the Pitman arm, you may encounter some made with bends and some without. Be aware of this when inspecting the suspension, because a good part may be unnecessarily condemned. You should be able to see an abnormal mark of some kind on the rod; or, if you know that a wheel might have been hit, then replace it. If toe is out of specifications and no other problems are found, then replace it and realign the vehicle to specifications.

Question #29

Answer A is wrong. Turning radius is set by the manufacture.

Answer B is wrong. The adjustment is to level the centerlink.

Answer C is wrong. Neither Technician is correct.

Answer D is correct. Neither Technician is correct. The adjustment is for assembly and manufacturing variables, and allows the center link to be level.

Question #30

Answer A is wrong. Technician B is also correct. The rack must be held while loosening the inner tie rod ends because they are threaded onto the rack.

Answer B is wrong. Technician A is also correct. Reduced articulation effort usually indicates worn tie rod ends.

Answer C is correct. Both Technicians are correct. Refer to specifications for any allowable movement. When these parts are worn, there is no service, except to replace them. If the rack is not held while loosening the inner tie rod ends, the rack can twist, causing internal damage.

Answer D is wrong. Both Technicians are correct.

Question #31

Answer A is wrong. A damaged power steering gear spool valve cannot transmit road shock. It controls fluid flow.

Answer B is correct. Only Technician B is correct. The damper is to absorb road induced movement or road shock. In order to check the steering damper on trucks, first inspect the damper for physical damage or signs of leaking. Replace the damper if any of these problems are present. The alteration of the suspension is another problem that may cause excessive road shock in the steering of 4x4 vehicles. If the suspension has been altered, the caster may be out of specifications toward the negative, causing the vehicle suspension not to respond normally when hitting bumps.

Answer C is wrong. Only Technician B is correct.

Answer D is wrong. Only Technician B is correct.

Question #32
Answer A is wrong. All power steering rack and pinion assemblies need adjustment.
Answer B is correct. Only Technician B is correct. All rack and pinion systems have several adjustments that need to be checked, regardless of power or manual systems.
Answer C is wrong. Only Technician B is correct.
Answer D is wrong. Only Technician B is correct.

Question #33
Answer A is correct. A worn strut rod bushing would not affect body sway while cornering. It would be noisy over bumps.
Answer B is wrong. Stabilizer bars control body roll; if they are weak, this would cause a problem.
Answer C is wrong. A worn stabilizer bar bushing would cause excessive play between the stabilizer bar and the bushing that would translate into body roll.
Answer D is wrong. If a stabilizer bar link is broken, the sway bar could not function properly. The automotive industry uses the term *returnability* to describe a driving situation when the steering wheel is turned in either direction and does not return back to a straight-ahead position. Several problems can cause this condition. First, check the steering for any binding of components, such as ball joints or tie rods. Then, check the rack and pinion preload on the pinion shaft. Finally, check to see if the caster is within specifications by performing an alignment.

Question #34
Answer A is correct. Only Technician A is correct. A binding and releasing action is a symptom of a faulty strut mount.
Answer B is wrong. Worn ball joints do not usually bind or steering chatter. A joint that is dry or frozen can.
Answer C is wrong. Only Technician A is correct.
Answer D is wrong. Only Technician A is correct.

Question #35
Answer A is correct. Only Technician A is correct. Worn control arm bushings will cause the vehicle alignment angles t change on acceleration. This can result in a pull on acceleration.
Answer B is wrong. Technician B is wrong. A power steering pump will not cause a pull. A power steering pump only provides fluid flow.
Only Technician A is correct.
Only Technician A is correct.

Question #36
Answer A is wrong. If a safety stand was placed under the chassis, the ball joint would still be loaded.
Answer B is wrong. Disconnecting the shock would not unload the ball joint.
Answer C is wrong. Disconnecting the sway bar link bushings does not unload the ball joint.
Answer D is correct. Until the safety stand is placed under the lower control arm, the spring is loading the ball joint. With the stand in place, the load is transferred to the stand (in place of the tire) where accurate ball joint inspection can be made.

Question #37
Answer A is correct. A bent steering arm (not to be confused with a Pitman arm) can change the turn radius, resulting in a change in the toe-out on turning distance. The tire that is making the larger arc will usually cause the squeal. A vehicle that has excessive tire squeal on turns is caused by less-than-specified toe-out on turns. This is usually caused by a bent tie rod on one side or the other. This should be the first thing to look for. The other problem that may cause this is a loose or worn tie-rod.
Answer B is wrong. Caster would not make the tire squeal when cornering.
Answer C is wrong. Stabilizer bar bushings have no effect on the steering geometry.
Answer D is wrong. Struts would have no effect on the steering geometry.

Question #38
Answer A is correct. The tool shown is used to compress the coil springs on a long-and-short arm suspension system. Working with coil springs can be dangerous. When removing or installing the front coil springs, the proper tool makes the job easier and safer. The spring tool needs an open hole through the center of the lower control arm and spring; so the shock needs to be removed. Insert the tool through the center of the lower control arm and through the spring and attach the two hooks onto the spring. Loosen the long center bolt on the tool until you can attach the bottom attachment to bolt. Then tighten the bolt on the tool, making sure it stays centered. (If you have any problems, loosen and readjust the bolt and restart the procedure.) As the spring is being compressed, stay clear in case the tool slips or breaks. Compress until you can install or remove the spring as needed and loosen the tool.
Answer B is wrong. There is no tool used to position the shock absorber.
Answer C is wrong. The ball joint movement is measured with a dial indicator.
Answer D is wrong. Lower control arm bushing wear is simply visually inspected.

Question #39
Answer A is wrong. Misadjusted tie-rod ends will affect toe.
Answer B is correct. Toe is an adjustment of linkage not the steering gearbox.
Answer C is wrong. A bent center link will affect toe. It would cause excessive toe-in.
Answer D is wrong. A bent Pitman arm will affect toe.

Question #40
Answer A is wrong. To properly align the castellation nut, torque the nut. Then tighten the nut to align the cotter key hole with nut castellations.
Answer B is correct. Only Technician B is correct. All steering and suspension bolts must be torqued following manufacturer's specifications to prevent loosening up and premature part failure. When servicing the ball joints, always torque the castle nut to specifications and check to see if the cotter pin hole is aligned. If the hole is not aligned, tighten the nut until you can insert the cotter pin into the castle nut; secure it by spreading the pin open after pushing it through the hole. Never loosen the castle nut to align the hole. This will take away the torque, and the ball joint shaft would come loose in the spindle hole.
Answer C is wrong. Only Technician B is correct.
Answer D is wrong. Only Technician B is correct.

Question #41
Answer A is wrong. Technician B is also correct. Oil-soaked mounting bushings can let the rack shift on the frame and cause the noise described.
Answer B is wrong. Technician A is also correct. Dried-out and loose mounting bushings can let the rack shift on the frame and cause the noise described.
Answer C is correct. Both Technicians are correct.
Answer D is wrong. Both Technicians are correct.

Question #42
Answer A is wrong. A broken center bolt in the left rear spring causes rear axle offset and this problem may cause steering pull to the right.
Answer B is wrong. Because the steering pulls to the side with the least positive caster, excessive positive caster on the left front wheel may cause steering pull to the right.
Answer C is wrong. More positive camber on the right front wheel compared to the left front wheel may cause steering pull to the right.
Answer D is correct. In this question the requested answer does not result in steering pull to the right. Excessive front wheel toe-in causes feathered tire wear, but this problem does not affect steering pull.

Question #43
Answer A is wrong. Sway bars are also called stabilizer bars.
Answer B is wrong. The sway bar *does* pivot in its mounts.
Answer C is correct. The lower control arms are mounted to the frame directly through rubber mount bushings.
Answer D is wrong. The sway bar *does* transfer movement from one side to the other to stabilize or level the vehicle.

Question #44
Answer A is wrong. Linear rate springs do not have a difference in space between the coils.
Answer B is correct. Coil spacing varies on variable-rate springs.
Answer C is wrong. Heavy-duty springs have thick coils or an additional helper spring in the center.
Answer D is wrong. The spring shown is not an adjustable coil spring. There is no such thing.

Question #45
Answer A is wrong. Technician B is also correct. Tire conicity definitely can cause the vehicle to pull.
Answer B is wrong. Technician A is also correct. Different tread patterns on the front tires can cause the vehicle to pull.
Answer C is correct. Both Technicians are correct. When diagnosing steering pull, check that the tires are of equal tread depth and of the same type and size. Next, check that tire pressure is set to specifications. These two simple checks will fix a majority of pull problems. It is best to replace tires in sets, however if they are replaced in pairs then, keep them on the same axle, either front or back, to keep the vehicle from pulling to one side or the other.
Answer D is wrong. Both Technicians are correct.

Question #46
Answer A is wrong. A broken center bolt in the leaf spring could cause the alignment angles to vary, causing a pull.
Answer B is wrong. The vehicle will lead to the side with the least amount of caster.
Answer C is wrong. A pull will occur if the tire pressures are not set properly.
Answer D is correct. Excessive toe-in causes feathered tire wear, but this problem does not affect steering pull.

Question #47
Answer A is wrong. Technician B is also correct.
Answer B is wrong. Technician A is also correct.
Answer C is correct. Both Technician A and B are correct. Jounce bumpers are located in different positions on different vehicles. Sometimes they are mounted on the frame, other times they are on the MacPherson strut rods.
Answer D is wrong. Both Technicians are correct.

Question #48
Answer A is wrong. Technician B is also correct. Some cradles have an alignment hole that must be matched to a corresponding hole in the chassis.
Answer B is wrong. Technician A is also correct. Several measurement points usually exist to determine cradle position and integrity.
Answer C is correct. Both Technicians are correct. Various manufacturers have published measurement procedures and specifications. If adjustments cannot be made within normal travel limits, it will be necessary to perform cradle measurements or alignment. Some cradles may already have an alignment dowel or hole for aligning it; no adjustment is needed on these. Cradles that require aligning have specifications and points to measure to so that the cradle can be centered. If you find that you cannot get it within specifications, then the frame of the vehicle may be bent—a body shop may have to check the frame for straightness. After the cradle is in correctly, a complete four-wheel alignment should be done.
Answer D is wrong. Both Technicians are correct.

Question #49
Answer A is wrong. Since the rear wheels are misaligned, it would not cause the steering to wander.
Answer B is correct. Because the rear wheels are offset, the thrust line will be off. This will result in a steering wheel that will not be centered correctly.
Answer C is wrong. Since the front suspension is aligned, returnability would not be affected.
Answer D is wrong. The pull would occur to the right.

Question #50

Answer A is correct. The length of the tie rod affects the rear toe only.

Answer B is wrong. This tie rod does not affect camber.

Answer C is wrong. The longer-than-normal tie rod would cause wear on the inside edge of the tire.

Answer D is wrong. Excessive toe-out on the left rear wheel will not cause a pull to the right.

Question #51

Answer A is wrong. The sensor is not normally replaced during a tire replacement. Normal sensor life is anticipated to be near 10 years. Most vehicles will need a set of tires long before then.

Answer B is correct. Only Technician B is correct. When the tires are rotated the computer must "learn" which sensor is at which position. This is accomplished in different ways by different manufacturers.

Answer C is wrong. Only Technician B is correct.

Answer D is wrong. Only Technician B is correct.

Question #52

Answer A is wrong. Technician B is also correct.

Answer B is wrong. Technician A is also correct.

Answer C is correct. Both Technicians are correct. If an air-assisted shock absorber slowly loses pressure, it may need replacement.

Answer D is wrong. Both Technicians are correct.

Question #53

Answer A is wrong. Tapered roller bearings require a very slight preload to then loosening to make the rollers roll properly.

Answer B is wrong. Tightening the spindle nut as much as possible by hand would be over tightening the nut and the bearings and cause premature bearing failure.

Answer C is wrong. Neither Technician is correct.

Answer D is correct. Neither Technician is correct. The bearings should be preloaded; then the end-play is set to .001"-.005".

Question #54

Answer A is wrong. It is not possible to remove the steering gear first; there are other components that have to be removed first, such as the pitman arm, steering shaft, and retaining bolts.

Answer B is wrong. This is not the easiest accessible component to disconnect.

Answer C is wrong. The steering-gear-to-frame mounts should be the last thing to be removed.

Answer D is correct. Disconnecting the battery cable will prevent problems while removing the steering gear, especially if working in the vicinity of the starter motor, or other "live" feeds. Removal of the manual steering gear should start with disconnection of the battery. Center the steering wheel and secure to avoid damage to the clock spring in the column. Now raise and support the vehicle and remove both front wheels. Disconnect the pitman arm joint and steering gear shaft coupling. From this point, you will need to remove the mount bolts to remove the steering gear. On some vehicles, you may also need to lower the sub-frame. Once the mount bolts are removed, the gear will come out.

Question #55

Answer A is correct. The system must be active in order for the scan tester to read data.

Answer B is wrong. Damage to the air suspension springs can occur if the switch is not turned off.

Answer C is wrong. When hoisting the vehicle, the air suspension switch must be turned off.

Answer D is wrong. The air suspension should be turned off any time the vehicle is towed.

Question #56

Answer A is correct. A loose cradle could cause the vehicle alignment angles to change upon acceleration. This could cause pull.

Answer B is wrong. A loose power steering belt would cause a squealing noise. It would not cause a pull.

Answer C is wrong. Only answer A is correct.

Answer D is wrong. Only answer A is correct.

Question #57
Answer A is wrong. Technician A is incorrect. The lamp has nothing to do with power steering fluid service intervals.
Answer B is wrong. Technician B is incorrect. The lamp has nothing to do with service intervals on the power steering pump belt.
Answer C is wrong. Neither Technician is correct.
Answer D is correct. Neither Technician is correct. The steering wheel lamp, which is fond on many new vehicles, indicates that the computer has identified a problem with the electronically controlled steering. A scan tool should be connected and the system scanned for trouble codes.

Question #58
Answer A is wrong. The pump shaft does not pass through the joint between the reservoir and the pump housing. It would be leaking out of the pump shaft seal.
Answer B is wrong. A restricted return hose would not cause the leak described in the question. It would cause excessive pressure in the rack assembly.
Answer C is wrong. Neither Technician is correct.
Answer D is correct. Neither Technician is correct. An inspection of the housing and seals may be in order. The power steering pump with an integral-type reservoir has an O-ring seal that can be replaced as needed, such as when leaking. To replace the seal, remove the power steering pump, as stated in the service manual Drain all fluid out of the pump and take a fluid sample to check for metal particles, which indicate that there may internal damage from running the pump low on fluid. Next, remove the housing by removing any bolts that secure the housing to the pump; then carefully separate the housing from the pump and note condition of seal. Clean the pump and reservoir; replace all seals, reassemble, and reinstall pump assembly. Fill and bleed the system; check for leaks. A pump pressure test should be done if you suspect a problem.

Question #59
Answer A is correct. Only Technician A is correct. This is an input signal; used to increase idle to compensate for power steering load.
Answer B is wrong. Would be "0" volts.
Answer C is wrong. Only Technician A is correct.
Answer D is wrong. Only Technician A is correct.

Question #60
Answer A is wrong. Usually if the steering gear requires adjustment, it is not because the vehicle experiences bump steer.
Answer B is correct. Only Technician B is correct. In this situation, the bent pitman arm causes unequal toe changes that occur as the suspension travels through jounce and rebound, resulting in sudden veering. Bump steer on a vehicle can cause loss of control, if serious enough, and should be corrected. When diagnosing bump steer, first verify this by conducting a road test. Then look for obvious signs of worn or loose suspension components. More often than not, it's a lower control arm bushing that causes this problem. Thoroughly check and replace defective parts as necessary. Then realign the vehicle to specifications and road test to verify the repair.
Answer C is wrong. Only Technician B is correct.
Answer D is wrong. Only Technician B is correct.

Question #61
Answer A is wrong. Technician B is also correct.
Answer B is wrong. Technician A is also correct.
Answer C is correct. Both Technicians are correct. Before an alignment is performed, check tire pressure, road test and do a complete suspension and steering inspection must be done.
Answer D is wrong. Both Technicians are correct.

Question #62
Answer A is wrong. If the steering gear were adjusted too tight, the steering would not return to center from a turn, and would be hard to steer through center.
Answer B is correct. Only Technician B is correct. Bump steer occurs when the tie rods are not parallel to the lower control arms. This condition may be caused by a bent pitman arm or a loose or improperly adjusted idler arm.
Answer C is wrong. Only Technician B is correct.
Answer D is wrong. Only Technician B is correct.

Question #63
Answer A is wrong. Technician B is also correct.
Answer B is wrong. Technician A is also correct.
Answer C is correct. Both Technicians are correct. The rear struts dampen road induced vehicle movement and maintain correct ride height. Ride height set too low can also cause bottoming out. When looking into the cause of a vehicle bottoming out, first find out when this problem occurs and if the vehicle is being overloaded. To verify the complaint, you may need to road test the vehicle with the addition of sand bags to simulate passengers in the vehicle. To eliminate this complaint, the springs, shocks, and struts are the most common items that require fixing.
Answer D is wrong. Both Technicians are correct.

Question #64
Answer A is wrong. Technician B is also correct.
Answer B is wrong. Technician A is also correct.
Answer is C is correct. Both Technicians are correct. Incorrect ride height may cause tires to wear even after caster, camber and toe are set to specification. If the suspension is operating outside of it's designed ride height it may experience extreme camber and toe changes in bump or rebound which will cause tire wear. Also, incorrect ride height may change the driveshaft working angle causing a vibration from the drive train in both front and rear wheel drive.
Answer D is wrong. Both Technicians are correct.

Question #65
Answer A is wrong. Technician B is also correct.
Answer B is wrong. Technician A is also correct.
Answer C is correct. Both Technicians are correct. Worn jounce bumpers are usually caused by the suspension traveling too far in the compressed direction. A low ride height setting can cause this as well as shock absorbers being worn.
Answer D is wrong. Both Technicians are correct.

Question #66
Answer A is correct. A vehicle will pull to the side with the least amount of caster. When diagnosing pulls while trying to drive straight, look at the tires and check for excessive road force induced pull from internal tire damage. Next, check that the brakes are not binding from a seized brake caliper piston or mounting. Also look at ride height to be out of specifications. Check the wheel bearings when inspecting the brakes, and check the alignment for excessive cross caster.
Answer B is wrong. A vehicle will pull to the side with the least amount of caster.
Answer C is wrong. Excessive toe-in will not cause a vehicle to pull to one side.
Answer D is wrong. Improper toe-out on the right front wheel while turning would not cause the vehicle to pull.

Question #67
Answer A is correct. This is a way of testing for worn shocks. When checking the shocks, inspect them with no excessive weight in the vehicle. Check the shocks for any leakage of fluid. Inspect the shock mounting and bushing for looseness.
Answer B is wrong. Coil springs do not dampen oscillations.
Answer C is wrong. A broken spring insulator would only cause noises when the suspension is jounced.
Answer D is wrong. Stabilizer bars only control body roll; a worn bushing would not affect the jounce ability.

Question #68
Answer A is correct. Only Technician A is correct. SAI problems affect camber and can prevent accurate camber adjustment.
Answer B is wrong. Elongating the adjustment holes are not an acceptable method of adjustment on an SLA suspension system.
Answer C is wrong. Only Technician A is correct.
Answer D is wrong. Only Technician A is correct.

Question #69
Answer A is correct. This aids in removal and installation to reduce binding and prevent tearing the tire bead.
Answer B is wrong. This may damage the beads.
Answer C is wrong. Never exceed 50 psi.
Answer D is wrong. Use appropriate tire machine equipment to avoid damage to rim or tire.

Question #70
Answer A is wrong. Technician B is also correct.
Answer B is wrong. Technician A is also correct.
Answer C is correct. Both Technicians are correct. The shim may have been installed to compensate for something other than the spindle and should be reinstalled when replacing the spindle. Anytime a steering or suspension component is replaced, a complete four-wheel alignment needs to be done.
Answer D is wrong. Both Technicians are correct.

Question #71
Answer A is correct. Camber and toe is read with wheels straight ahead, not caster.
Answer B is wrong. The front wheels must be turned left and right to measure caster.
Answer C is wrong. The brakes are applied so that the adjustments do not change while the retaining bolts are loose.
Answer D is wrong. Before checking caster, the suspension system must be settled by jouncing the vehicle's front suspension.

Question #72
Answer A is wrong. Replacing the struts will not necessarily correct caster unless struts are bent.
Answer B is wrong. A technician should never heat and bend a suspension component.
Answer C is correct. Cradle, or subframe, misalignment is a common cause of incorrect caster.
Answer D is wrong. Thrust angle does not affect caster. It affects wheel tracking.

Question #73
Answer A is wrong. Technician B is also correct.
Answer B is wrong. Technician A is also correct.
Answer C is correct. Both Technicians are correct Memory steer is when the vehicle fails to return straight, or wants to wander to the direction it was last turned. A binding or seized strut upper-mount bearing would cause this problem.
Answer D is wrong. Both Technicians are correct.

Question #74
Answer A is wrong. The steering may not be centered even though the steering wheel is centered.
Answer B is correct. Only Technician B is correct. The steering wheel needs to be centered in relationship to the equal number of turns from center to lock, left and right. Other adjustments are also based off of this reference such as steering gear preload.
Answer C is wrong. Only Technician B is correct.
Answer D is wrong. Only Technician B is correct.

Question #75
Answer A is wrong. Technician B is also correct. SAI correction may require replacement of damaged parts.
Answer B is wrong. Technician A is also correct. SAI problems may require correction before camber can be adjusted.
Answer C is correct. Both Technicians are correct.
Answer D is wrong. Both Technicians are correct.

Question #76
Answer A is wrong. Included angle is always a concern and can cause tire wear. It's an indication of worn or bent suspension components.
Answer B is correct. Only Technician B is correct. Included angle will effect vehicle stability and handling.
Answer C is wrong. Only Technician B is correct.
Answer D is wrong. Only Technician B is correct.

Question #77
Answer A is wrong. Placing the shim on the front spindle bolts would increase the left rear wheel toe-out.
Answer B is wrong. If the rear wheel(s) are offset to the left, it will cause the steering wheel to be off center.
Answer C is wrong. Neither Technician is correct.
Answer D is correct. Neither Technician is correct. When looking into adjusting the toe on the rear of a front wheel drive vehicle, there are several different methods to use, depending on the manufacturer and vehicle suspension type. When using shims behind the spindle and the spindle mounting surface, there are two types that can be used: full-circle shims or part-circle shims. The full circle has a thickness that is different around the circumference of the circle than the part circle. Most of the time, they are made of plastic covering all four attaching spindle bolts. The semicircle shim covers only the front two attaching spindle bolts. Both types of shims will cause the front of the spindle to increase or decrease toe.

Question #78
Answer A is wrong. Camber does not feather the tread. Excessive positive camber will cause wear on the outside tread area of the tire.
Answer B is wrong. Caster angle does not cause tire wear.
Answer C is wrong. Setback does not cause tire wear.
Answer D is correct. Toe adjustment not as specs will cause the tire to "scrub" sideways.

Question #79
Answer A is wrong. Worn ball joints do not usually cause hard steering.
Answer B is correct. Only Technician B is correct. Tight rack bearings increase drag on the rack and can cause hard steering, or binding.
Answer C is wrong. Only Technician B is correct.
Answer D is wrong. Only Technician B is correct.

Question #80
Answer A is wrong. Overinflation causes wear on the center of the tire tread.
Answer B is wrong. Underinflation causes wear on the edges of the tread.
Answer C is correct. The question asks for the statement that is not true. Tire pressures should be adjusted when the tires are cool.
Answer D is wrong. Underinflation may also cause wheel damage.

Question #81
Answer A is wrong. Technician A is also correct. A defective tire can cause this.
Answer B is wrong. Technician B is also correct. A bent wheel can cause this problem.
Answer C is correct. Both Technicians are correct. When encountering a ride problem that starts at very low speeds, inspect the tires. Look for signs of damage from impacting an object like a curb (most common). Then check the tires and rims on a wheel balance for rim run out and for broken steel belts inside. If all of these items check out okay, then look into a suspension component problem, starting with raising the vehicle and looking for loose items.
Answer D is wrong. Both Technicians are correct.

Question #82
Answer A is correct. A bent rear spindle can cause the camber angle to change, which could result in inside tread wear on the right rear tire.
Answer B is wrong. We normally do not have caster angles on the rear of a vehicle. Caster is a steering angle. Additionally, caster is not usually considered to be a tire wear angle.
Answer C is wrong. Excessive positive camber can cause wear. However, it will wear the outside tread, not the inside.
Answer D is wrong. Worn front struts will not cause rear tire wear.

Question #83
Answer A is wrong. Technician B is also correct.
Answer B is wrong. Technician A is also correct.
Answer C is correct. Both Technicians are correct. At regular recommended intervals, rotating tires can help in wearing all the tires at an equal rate. Some tires may need to be kept on the same side of the vehicle (directional rotation type), while others can be moved side to side. Both technicians are correct.
Answer D is wrong. Both Technicians are correct.

Answers to the Test Questions for the Additional Test Questions Section 6

1. B	19. A	37. C	55. C
2. C	20. A	38. C	56. A
3. C	21. A	39. B	57. A
4. B	22. A	40. A	58. A
5. A	23. C	41. C	59. C
6. D	24. C	42. A	60. D
7. C	25. A	43. C	61. C
8. C	26. A	44. C	62. B
9. A	27. A	45. A	63. A
10. B	28. C	46. B	64. D
11. B	29. C	47. A	65. B
12. C	30. A	48. A	66. A
13. C	31. B	49. A	67. A
14. D	32. C	50. B	68. B
15. C	33. B	51. D	69. C
16. B	34. D	52. B	70. A
17. A	35. B	53. C	71. C
18. D	36. D	54. D	72. C

Explanations to the Answers for the Additional Test Questions Section 6

Question #1
Answer A is wrong. Running a tire under its specified pressure can damage the tire and is not recommended.
Answer B is correct. Only Technician B is correct. Tire lead is a common cause of steering pull and sometimes can be corrected by rotating tires.
Answer C is wrong. Only Technician B is correct.
Answer D is wrong. Only Technician B is correct.

Question #2
Answer A is wrong. Technician B is also correct.
Answer B is wrong. Technician A is also correct.
Answer C is correct. Both Technicians are correct. Directional tires and wheels need to be kept on the correct side of the vehicle.
Answer D is wrong. Both Technicians are correct.

Question #3
Answer A is wrong. Technician B is also correct. Movement can cause excessive setback.
Answer B is wrong. Technician A is also correct. Cradle alignment could fix setback.
Answer C is correct. Both Technicians are correct. If the cradle cannot be centered or repositioned per manufacturer procedures, then specific measurements must be made, as per the points specified. If it is damaged, it will need to be replaced.
Answer D is wrong. Both Technicians are correct.

Question #4
Answer A is wrong. This is generally a waste of time. While it is nice to get them close to the same position, this can only be done during alignment.
Answer B is correct. This will insure that the steering column and gear are phased correctly. Always follow appropriate safety precautions. Then, to keep from damaging the clock spring in the column, note the location of the steering shaft to steering gear pinion after centering the wheel and securing the steering wheel. When removing the tie rod ends that are being transferred from the old to the new rack and pinion, measure the installed distance so that they will only require minimal adjustment during the alignment when putting them on the new rack and pinion. Before installation, turn the rack and pinion gear pinion shaft all the way left to right; then turn back 1/2 of the total turns made; do so to center the rack and its travel left to right to match the steering wheel. Use the service manual as needed, to install and secure the rack into the vehicle.
Answer C is wrong. In most applications, the stabilizer links will not have to be disconnected and they have no relationship to steering.
Answer D is wrong. The steering gear mounting bolts are in fixed positions so they do not require marks to return them to original position.

Question #5
Answer A is correct. Tire balance does not affect steering wheel return.
Answer B is wrong. When a tire is unbalanced, it begins to bounce and does not contact the road surface evenly, causing cupped tire wear.
Answer C is wrong. When the tire is unbalanced, the vibrations can be transmitted through the steering linkage into the steering wheel.
Answer D is wrong. Vibrations that are caused from unbalanced tires can cause premature shock/strut wear.

Question #6
Answer A is wrong. Loose worm bearing preload would cause the steering effort to be minimal, not excessive.
Answer B is wrong. An overfilled steering gear would not affect steering effort.
Answer C is wrong. The more positive caster a vehicle has the harder it will be to turn.
Answer D is correct. A conventional recirculating ball steering gear will demonstrate excessive steering effort if the sector adjustment is too tight. Over-tightening can damage the gears.

Question #7
Answer A is wrong. Technician B is also correct.
Answer B is wrong. Technician A is also correct.
Answer C is correct. Both Technicians are correct. There are no serviceable parts in this center link. Worn joints, elongated holes, or bends require replacing the center link.
Answer D is wrong. Both Technicians are correct.

Question #8
Answer A is wrong. Worn ball joints could cause excessive play in the steering, but not excessive effort.
Answer B is wrong. Weak front springs would affect the ride height of the vehicle.
Answer C is correct. A seized idler arm will add resistance and increase steering effort.
Answer D is wrong. A weak stabilizer bar would only affect the ride quality of the vehicle on turns. When looking at increased or excessive steering effort, a full inspection of all steering and suspension components needs to be done to check for excessive wear or binding. Look for torn or leaking joint grease boots that may have lost their lubricant or let in dirt and water to compromise them. Don't forget to always check for low tire pressures.

Question #9
Answer A is correct. Only Technician A is correct. With a live axle you will usually find that a thrust angle problem will result in toe in on one side and a nearly equal amount of toe out on the other side.
Answer B is wrong. Either condition will cause the steering wheel to be off-center but a live axle application does not have a toe adjustment. If toe is out of spec and cannot be corrected by moving the axle housing around the housing is bent.
Answer C is wrong. Only Technician A is correct.
Answer D is wrong. Only Technician A is correct.

Question #10
Answer A is wrong. Poor returnability would not be caused by a loose rack bearing adjustment.
Answer B is correct. Insufficient caster on a manual rack and pinion vehicle may cause poor returnability. Positive caster would help steering returnability.
Answer C is wrong. Loose steering gear mounting bolts would cause a clunk when cornering. That would not usually cause poor returnability.
Answer D is wrong. Excessive positive camber would only cause tire wear and may cause a pull to one side.

Question #11
Answer A is wrong. A front-wheel alignment would not adjust the rear of vehicle.
Answer B is correct. Anytime a steering or suspension component is replaced a four-wheel alignment needs to be performed, starting with rear wheels.
Answer C is wrong. You would replace the struts only if they are worn or defective.
Answer D is wrong. There would be no need to service the steering hydraulic system.

Question #12
Answer A is wrong. Technician B is also correct.
Answer B is wrong. Technician A is also correct.
Answer C is correct. Both Technicians are correct. Technician A is right. After repair of the power steering rack both rack bellows boots should be replaced because the fluid can soften them and the equalizer tube will transfer fluid from side-to-side. Technician B is also correct. The rack and pinion gear must be repaired or replaced.
Answer D is wrong. Both Technicians are correct.

Question #13
Answer A is wrong. Technician B is also correct. A restricted high-pressure hose can reduce pressure at the steering gear even when the pump is developing full pressure due to collapsed inner hose lining or kinked line.
Answer B is wrong. Technician A is also correct. A defective power steering gear can cause excessive steering effort due to a faulty control valve even with proper hydraulic pressure.
Answer C is correct. Both Technicians are correct. When diagnosing a problem with steering effort, the problem can be narrowed down by starting with a thorough road test. Then inspect steering components for problems. Remove the pressure line from the pump, attach a tester, and check for proper pressures. If this passes, do the same test at the other end of the pressure line where it connects to the rack and pinion; this will tell you if the line is collapsed inside. If these tests are normal, inspect the return hose for a blockage to determine if the problem is inside the rack and pinion assembly. Inspect and service it as necessary.
Answer D is wrong. Both Technicians are correct.

Question #14
Answer A is wrong. Rear wheels should be adjusted before trying to adjust the front wheels.
Answer B is wrong. Ride height would not affect thrust angle.
Answer C is wrong. Rotating the tires will not correct thrust angle.
Answer D is correct. A four-wheel alignment is needed, starting with the rear wheels.

Question #15
Answer A is wrong. Technician B is also correct.
Answer B is wrong. Technician A is also correct.
Answer C is correct. Both Technicians are correct. Any component that has a provision for lubrication must be greased as prescribed in the maintenance schedule. Removing the tie rod eliminates steering linkage drag while checking the kingpins.
Answer D is wrong. Both Technicians are correct.

Question #16
Answer A is wrong. This would be done after the rack unit is removed.
Answer B is correct. For safety reasons, you must disable the air bag system following manufacturer's recommendations. Also locking the steering wheel in place will insure no damage is done to the clock spring (spiral cable) when the steering column is disconnected from the rack and pinion.
Answer C is wrong. The power steering lines must be removed to remove the rack; however they are not the first item to be removed.
Answer D is wrong. Although the power steering reservoir must be drained, it is not the first step.

Question #17
Answer A is correct. Only Technician A is correct. Most vehicles have a pressed on pulley that can be moved into alignment.
Answer B is wrong. Using shims to align the power steering pulley is not an approved practice.
Answer C is wrong. Only Technician A is correct.
Answer D is wrong. Only Technician A is correct.

Question #18
Answer A is wrong. Sway bars can cause popping noises if not seated correctly.
Answer B is wrong. Stiff springs would cause a loaded vehicle to ride higher than normal.
Answer C is wrong. Neither Technician is correct.
Answer D is correct. Neither Technician is correct. This problem would be caused by a driveshaft working angle problem that occurs when there is a load in the rear. It is possible that inadequate rear springs might allow this change to occur. Springs with too much spring rate or sway bars would not cause this kind of problem.

Question #19
Answer A is correct. Only Technician A is correct. Oil leaking from the shock absorber is a good indication that the shock is defective and should be replaced.
Answer B is wrong.. Although the shock passes a bounce test, if the shock is leaking, it eventually will not be able to dampen suspension oscillations.
Answer C is wrong. Only Technician A is correct.
Answer D is wrong. Only Technician A is correct.

Question #20
Answer A is correct. Only Technician A is correct. During braking, a worn strut bushing would cause a rapid caster change.
Answer B is wrong. Negative caster would cause a pull, regardless of braking action.
Answer C is wrong. Only Technician A is correct.
Answer D is wrong. Only Technician A is correct.

Question #21
Answer A is correct. Only Technician A is correct. The steering wheel should not move because it can damage the air bag system clock spring.
Answer B is wrong. The clock spring connector is not removed.
Answer C is wrong. Only Technician A is correct.
Answer D is wrong. Only Technician A is correct.

Question #22
Answer A is correct. Shock absorbers do not affect ride height. They affect ride smoothness and steering stability.
Answer B is wrong. A broken coil spring will alter ride height and cause the vehicle to lean to one side.
Answer C is wrong. A misadjusted torsion bar will alter ride height and cause the vehicle to lean to one side.
Answer D is wrong. A weak or sagging leaf spring will alter ride height and cause the vehicle to lean to one side.

Question #23
Answer A is wrong. Both Technicians are correct.
Answer B is wrong. Both Technicians are correct.
Answer C is correct. Both Technicians are correct. The clamp must be positioned in such a way so that it squeezes around the solid portion of the sleeve equally and not at or on the slot. The clamps are usually rotated so that the open end is down, or otherwise positioned so it will not come in contact with anything during travel while turning.
Answer D is wrong. Both Technicians are correct.

Question #24
Answer A is wrong. A bent center link would affect the alignment more than the steering effort.
Answer B is wrong. Technician A is also correct.
Answer C is correct. Both Technicians are correct. Although within specs, it is best to adjust to the preferred specification where possible.
Answer D is wrong. Both Technicians are correct.

Question #25
Answer A is correct. Only Technician A is correct. Worn control arm bushings can cause the caster angle to be out of limits.
Answer B is wrong. Caster angles do not cause abnormal tire wear patterns.
Answer C is wrong. Only Technician A is correct.
Answer D is wrong. Only Technician A is correct.

Question #26
Answer A is correct. Only Technician A is correct. Preload adjustments are performed using various methods, depending on the manufacturer.
Answer B is wrong. The steering gear does not have to be drained to be adjusted.
Answer C is wrong. Only Technician A is correct.
Answer D is wrong. Only Technician A is correct.

Question #27
Answer A is correct. Only Technician A is correct. A leaking or dented steering damper will not work properly. A steering damper is used to lessen road shock through the steering.
Answer B is wrong. High-speed vibration is not likely to be caused by a faulty steering damper. Other possible causes are more likely such as worn components or tire problems.
Answer C is wrong. Only Technician A is correct.
Answer D is wrong. Only Technician A is correct.

Question #28
Answer A is wrong. Technician B is also correct.
Answer B is wrong. Technician A is also correct.
Answer C is correct. Both Technicians are correct. Eye bushings can wear due to age and prolonged twisting action associated with normal use. In addition, the separators can shift or wear. Either condition can cause irregular spring movement.
Answer D is wrong. Both Technicians are correct.

Question #29
Answer A is wrong. Technician B is also correct.
Answer B is wrong. Technician A is also correct.
Answer C is correct. Both Technicians are correct. Worn strut bushings can cause a caster angle change, resulting in a pull problem.
Answer D is wrong. Both Technicians are correct.

Question #30
Answer A is correct. Only Technician A is correct. A soap solution will bubble when a leak is found.
Answer B is wrong. While the vehicle is standing overnight, it will probably leak down, and the technician still will not know where the leak is.
Answer C is wrong. Only Technician A is correct.
Answer D is wrong. Only Technician A is correct.

Question #31
Answer A is wrong. Excessive toe-out can only cause tire wear.
Answer B is correct. When a tire cord breaks the tire tends to squirm as it rotates which is perceived by the driver as a side to side oscillation (shimmy). The tire will not hold a balance either.
Answer C is wrong. Setback angles cannot cause shimmies or vibration. It may cause a pull or drift to one side or the other.
Answer D is wrong. Camber angles do not cause shimmy. It may cause pull to one side or the other and tire wear. To check for the cause of harsh ride, first road test to feel or hear the concern, then look at the quality, type, and size of tires on the vehicle. Check the tire pressure and check for excessive weight in the trunk. Make the necessary adjustments; then check the condition of the suspension components.

Question #32
Answer A is wrong. Technician B is also correct. The height sensor adjustment may have slipped.
Answer B is wrong. Technician A is also correct. The height sensor may be disconnected.
Answer C is correct. Both Technicians are correct.
Answer D is wrong. Both Technicians are correct.

Question #33
Answer A is wrong. The inner tie rod is attached to the rack shaft threaded end.
Answer B is correct. The part indicated is the rack outer seal bushing. The wrenches illustrate the proper way to remove the inner tie rod assembly. While servicing the inner tie rod, to protect the rack shaft from damage, use a protective shield, such as a rag. A nick or scar on the rack shaft would cause the rack side seal to leak around the rack shaft.
Answer C is wrong. This is not the rack piston ball servo.
Answer D is wrong. The inner tie rod is attached to the rack shaft threaded end.

Question #34
Answer A is wrong. A pump problem will also cause difficult steering while parking.
Answer B is wrong. A steering gear problem will affect steering all the time.
Answer C is wrong. Rotation sensor problems may cause the system to go to full steering-assist.
Answer D is correct. If a problem existed in the pump or gear, there would be no power steering during parking. The speed sensor sends vehicle speed information to the computer. Steering assist will decrease at higher speeds and increase at lower speeds.

Question #35
Answer A is wrong. Heat should not be applied to a torsion bar. It will cause part failure.
Answer B is correct. Tension is changed by turning an adjusting bolt to increase or decrease tension.
Answer C is wrong. Adding shims will not adjust the torsion bar.
Answer D is wrong. There are no brackets with offset.

Question #36
Answer A is wrong. A bent centerlink will cause toe to be out of adjustment.
Answer B is wrong. A loose power steering belt would make a squeal noise.
Answer C is wrong. Worn tie-rod ends would cause looseness in the steering.
Answer D is correct. Loose steering gear retaining bolts may cause the steering gear to shift on the frame while turning. This can cause a snapping noise.

Question #37
Answer A is wrong. Technician B is also correct.
Answer B is wrong. Technician A is also correct.
Answer C is correct. Both Technicians are correct. The movement of either or both the rod adjusting sleeves will cause the toe adjustment to change. It is possible to have correct toe angles and the steering wheel off center. The opposite is also true, whereby the steering wheel is centered. But, the toe angle is incorrect. The objective is to center the steering wheel, then adjust the toe as necessary using the tie rod sleeve(s).
Answer D is wrong. Both Technicians are correct.

Question #38
Answer A is wrong. Technician B is also correct.
Answer B is wrong. Technician A is also correct.
Answer C is correct. Both Technicians are correct. Regardless of the method, the load must be removed from the ball joint to accurately check for wear. If not, a worn ball joint may be missed or overlooked because the weight of the load will make it seem tight.
Answer D is wrong. Both Technicians are correct.

Question #39
Answer A is wrong. Camber adjustments can be made by adding shims.
Answer B is correct. It is never recommended to heat and/or bend a control arm. Doing so can alter the designed strength specifications and may cause metal fatigue and failure.
Answer C is wrong. Camber adjustments can be made by installing cams.
Answer D is wrong. Camber adjustments can be made by installing ball joint eccentrics.

Question #40
Answer A is correct. This is the correct and safe repair procedure.
Answer B is wrong. Disconnecting the ball joint with tension on the spring could cause personal injury.
Answer C is wrong. Bushings should never be lubricated with transmission oil.
Answer D is wrong. Disconnecting the outer tie-rod will not help in removing the lower control arm.

Question #41
Answer A is wrong. Stabilizer bars do not affect ride height. Stabilize bars maintain stability on turn and reduce body roll.
Answer B is wrong. Shock absorbers help dampen bounce, not stabilizer bars.
Answer C is wrong. Neither Technician is correct.
Answer D is correct. Neither Technician is correct.

Question #42
Answer A is correct. By changing the ride height, the alignment angles will be affected both in the front and in the rear suspensions. Most rear suspensions that have leaf-style springs cannot be adjusted. If out of alignment, they should be replaced. If you have a weak leaf spring or the vehicle leans to the one side, check the conditions of the spring support bushings for wear and replace, if needed. If the spring is suspected of being weak and sagging, replace it with a new or rebuilt one. Then, see if the left side and right side of the vehicle is even. If not, then it may be necessary to replace both of the springs to keep the vehicle even. Once the vehicle is leveled, recheck the alignment as the rear ride height may affect the front measurements.
Answer B is wrong. Rear tire wear would not be affected in this particular suspension design.
Answer C is wrong. The steering wheel free play is not affected by sagging rear springs.
Answer D is wrong. Sagging rear leaf springs do not directly affect front steering linkage.

Question #43
Answer A is wrong. Technician B is also correct. The spring must be fully compressed before loosening the strut rod nut.
Answer B is wrong. Technician A is also correct. The spring should be taped to protect it and to help keep the spring from slipping in the compressor.
Answer C is correct. Both Technicians are correct. When removing the complete strut assembly from a vehicle, follow all safety precautions. Removing a strut spring can be dangerous. Secure the strut assembly in the spring compressor and compress the spring only enough to take the tension off of the top mount assembly; then, remove the top mount nut. Replace components as needed; reassemble the strut assembly and torque the top nut to specifications.
Answer D is wrong. Both Technicians are right.

Question #44
Answer A is wrong. Technician B is also correct.
Answer B is wrong. Technician A is also correct.
Answer C is correct. Both Technicians are correct. If the bushing is in good condition, silicone lubricant can be used.
Answer D is wrong. Both Technicians are correct.

Question #45
Answer A is correct. Only Technician A is correct. The front adjustments are based on the rear settings. Therefore the rear adjustments should be done first.
Answer B is wrong. The rear wheels are always aligned before the front wheels on a four-wheel alignment.
Answer C is wrong. Only Technician A is correct.
Answer D is wrong. Only Technician A is correct.

Question #46
Answer A is wrong. Spring insulators should be placed on the springs or noise will result.
Answer B is correct. Variable rate springs are designed to deflect or compress more slowly as additional weight is added. The weight and compression ratio are inversely proportionate to the limit of the spring. The close convolutions are at the top, to give good ride characteristics with normal weight.
Answer C is wrong. Closed coils should be at the top.
Answer D is wrong. Springs should be placed with the ends about one half inch from their stops.

Question #47
Answer B is wrong. Realigning the vehicle will not repair the problem.
Answer C is wrong. There is no need to put maximum air pressure in the tires.
Answer D is wrong. Tightening the wheel nuts will not noticeably affect runout, or cause a pull problem.

Question #48
Answer A is correct. Only Technician A is correct. The total toe specification should be split left and right, for a total toe angle. This also helps ensure a centered steering wheel.
Answer B is wrong. Toe specifications are understood to be total toe unless specified left and right.
Answer C is wrong. Only Technician A is correct.
Answer D is wrong. Only Technician A is correct.

Question #49
Answer A is correct. This results in the symptom known as bump steer. Worn rack mounts would allow the rack to shift side to side. "Bump steer"; can be very dangerous and cause loss of control of the vehicle. Most of the time, bump steer problems can be contributed to problems with the rack and pinion mounting. It may be loose mounting bolts or bushings on which the rack is mounted. Though not as common a problem, loose sub-frame mountings can cause a shift or movement when hitting a bump.
Answer B is wrong. Caster adjustment would not affect the drivability of the vehicle if it hit a bump.
Answer C is wrong. Worn upper strut mounts would cause noises when cornering and increased steering effect.
Answer D is wrong. Worn front struts cause a floating ride quality.

Question #50
Answer A is wrong. The rack seal bypassing would cause a lack of power assist in both directions.
Answer B is correct. Only Technician B is correct. The spool valve controls the power assist and must maintain good seal between the housing and the spool. It is possible to lose one side assist with internal seal failure.
Answer C is wrong. Only Technician B is correct.
Answer D is wrong. Only Technician B is correct.

Question #51
Answer A is wrong. Excessive positive camber does not increase steering effort, and it has only a negligible effect on steering return.
Answer B is wrong. Reduced rear ride height has the effect of increasing positive caster on the front wheels, but it would take a very dramatic (unlikely) change to cause this customer concern. The customer would be to your shop to have his rear bumper replaced first.
Answer C is wrong. Neither Technician is correct.
Answer D is correct. Neither Technician is correct. Steering wheel poor return ability is almost always the result of excessive negative caster. The added weight toward the front from the negative caster also contributes to the increase in steering effort.

Question #52
Answer A is wrong. While inadequate caster (close to zero degrees) can cause decreased directional stability it is not a cause for bump steer.
Answer B is correct. Only Technician B is correct. Worn rack bushings can allow a bump to move the rack effectively changing the toe direction of the vehicle and the steering wheel to be misaligned. it is not a cause of bump steer.
Answer C is wrong. Only Technician B is correct.
Answer D is wrong. Only Technician B is correct.

Question #53
Answer A is wrong. An alignment will not fix a vibration problem.
Answer B is wrong. Shocks are for ride smoothness and to absorb road shock.
Answer C is correct. The problem described is a wheel balance problem. In most vehicles if the vibration can be felt in the steering wheel, the problem is with the front wheels. If the vibration can be felt in the seat, it is a rear wheel balance problem. It is a good idea to check shock operation after correcting the problem since wheel balance problems cause the shocks to overheat and wear out prematurely.
Answer D is wrong. An unbalanced drive shaft would most likely be felt throughout the vehicle.

Question #54
Answer A is wrong. A worn U-joint on the steering shaft can bind when worn.
Answer B is wrong. A loose belt can cause belt slippage and loss of power-assist.
Answer C is wrong. Binding upper strut mount bearing can cause excessive steering effort.
Answer D is correct. Worn rack and pinion bushings may cause a loose feel and wandering, but not excessive steering effort. Answer D is not correct. All of the other answers can cause excessive steering effort.

Question #55
Answer A is wrong. Technician B is also correct.
Answer B is wrong. Technician A is also correct.
Answer C is correct. Both Technicians are correct. Although within specs, it is best to adjust to the preferred specification where possible.
Answer D is wrong. Both Technicians are correct.

Question #56
Answer A is correct. When the camber is significantly different side to side with one going negative and the other positive, check for a cradle that has shifted. First, readjust the cradle in accordance with the shop manual specifications and adjustments. Then do four-wheel alignment. On some vehicles, alignment of the cradle cannot be obtained due to damage; the cradle may have to be replaced if it cannot be brought into alignment.
Answer B is wrong. Worn steering rack bushings will affect toe, not camber.
Answer C is wrong. Loose wheel bearings can cause incorrect camber reading but this is not the best answer for this particular question.
Answer D is wrong. Sagged springs can cause camber to change, but should affect both sides the same amount. The front suspension cradle can cause the camber to be out of specifications if the cradle is installed incorrectly or if the cradle was moved by an impact.

Question #57
Answer A is correct. A shorted switch sends a false signal and may result in higher than normal idle. It may also prohibit air conditioning operation.
Answer B is wrong. An open here would show normal in this condition.
Answer C is wrong. An open here would show normal in this condition.
Answer D is wrong. An open here would show normal in this condition.

Question #58
Answer A is correct. Only Technician A is correct. Conicity is a term for a type of tire wear that causes the tire to wear across the tread face in a cone shape. Vehicles that run positive or negative camber will develop this type of wear. Over time through rotation, the tires can wear in such a way that the worn tires bias the vehicle in one direction causing it to pull. This is a difficult symptom to diagnose without special tire balancing equipment.
Answer B is wrong. If a wheel or tire assembly was badly out of round, it would cause a violent vibration.
Answer C is wrong. Only Technician A is correct.
Answer D is wrong. Only Technician A is correct.

Question #59
Answer A is wrong. When a tire is overinflated, the center of the tread expands, causing wear in the center of the tire.
Answer B is wrong. When a tire is underinflated, the weight of the vehicle is positioned on the sidewall, causing wear to the sides.
Answer C is correct. Tire pressure should be adjusted when the tires are cold. The pressures should be adjusted to the car's recommended specifications; these are stated in the owners guide or on the vehicle safety data sticker in the door or glove box. The tire pressure on the side of the tire is the maximum pressure to which the tire can be safely inflated.
Answer D is wrong. Damage can occur to the tire and the wheel when tire inflation is below the specified pressure by the rim cutting into the tire's sidewall. Tires and tire pressure should be checked on a monthly basis. Prior to checking tires, they should be cold to the touch and have been driven less than 3 miles.

Question #60
Answer A is wrong. This is not a rear caster adjustment.
Answer B is wrong. Rear wheels do not have this provision on solid rear axles.
Answer C is wrong. This is not an ABS component.
Answer D is correct. Shims, in different thicknesses, are used for toe adjustment and sometimes camber correction.

Question #61
Answer A is wrong. Technician B is also correct.
Answer B is wrong. Technician A is also correct.
Answer C is correct. Both Technicians are correct. Any excessive runout will cause a vibration. The tires may need to be repositioned on the wheel, and rebalanced.
Answer D is wrong. Both Technicians are correct.

Question #62
Answer A is wrong. Ride height does not affect included angle but should be checked prior to alignment.
Answer B is correct. Camber angle is part of the included angle.
Answer C is wrong. Thrust angle does not affect included angle.
Answer D is wrong. The springs will not correct included angle.

Question #63
Answer A is correct. Only Technician A is correct. Camber is the inward and outward tilt of the wheel as viewed from the top of the tire. When looking for a camber problem, always remember that a vehicle will pull to the side that is most positive. Aligning a vehicle to within specifications is not the only thing to consider. For example, with a specification of + or − 6 degrees, it's often assumed that within these numbers the vehicle will go straight—this isn't always true. For instance, if you adjust a vehicle to have a +5 degree camber on the right side and a -5 degree camber on the left side, then you would be within specifications; however, you would have a cross caster of +10 degrees to the right. This is out of specifications and could cause a pull to the right side. For this reason, you must also look at cross-caster when adjusting caster.
Answer B is wrong. Moving the control arm forward will increase caster, not camber.
Answer C is wrong. Only Technician A is correct.
Answer D is wrong. Only Technician A is correct.

Question #64
Answer A is wrong. Caster does not wear tires.
Answer B is wrong. Toe-out would cause inside tire wear and scuffing.
Answer C is wrong. Neither Technician is correct.
Answer D is correct. Neither Technician is correct. Excessive tire wear on the outside tread is most likely a camber problem.

Question #65
Answer A is wrong. The front wheels must be aligned while in the straight-ahead position.
Answer B is correct. Toe is adjusted *after* caster and camber.
Answer C is wrong. While setting the toe, the steering wheel must be centered.
Answer D is wrong. Tie rod sleeves can become rusted or corroded; the use of a rod sleeve adjusting tool can help.

Question #66
Answer A is correct. After adjusting toe the lock nuts must be tightened to prevent movement of the shaft in the threads.
Answer B is wrong. No steering gear adjustment is needed after adjusting toe.
Answer C is wrong. The steering wheel should be centered before adjusting toe.
Answer D is wrong. There is no idler arm on a rack and pinion system.

Question #67
Answer A is correct. Only Technician A is correct. The bearing and race needs to be replaced as a unit due to wear or damage to one or the other.
Answer B is wrong. Not all bearings have failures at the same time.
Answer C is wrong. Only Technician A is correct.
Answer D is wrong. Only Technician A is correct.

Question #68
Answer A is wrong. The system does just the opposite; the steering assist is increased for parking maneuvers.
Answer B is correct. Only Technician B is correct. Reducing power assist at highway speeds gives the drive better road "feel" and decreases the chances of over-reaction at higher speeds.
Answer C is wrong. Only Technician B is correct.
Answer D is wrong. .Only Technician B is correct.

Question #69
Answer A is wrong. Technician B is also correct.
Answer B is wrong. Technician A is also correct.
Answer C is correct. Both Technicians are correct. Flexible couplings are used to isolate vibration in the steering column. If the coupling or steering shaft U-joints become loose or worn, loose steering or excessive play will result.
Answer D is wrong. Both Technicians are correct.

Question #70
Answer A is correct. Only Technician A is correct. The strut or shock absorber is designed to smooth out road-induced movement and body bounce.
Answer B is wrong. Coil springs do not dampen oscillations; that is the job of the strut or shock absorber.
Answer C is wrong. Only Technician A is correct.
Answer D is wrong. Only Technician A is correct.

Question #71
Answer A is wrong. The sector shaft should be checked for wear.
Answer B is wrong. The gear housing should be checked for wear.
Answer C is correct. The rack piston is found on rack and pinion steering only.
Answer D is wrong. The Pitman arm should be checked for wear.

Question #72
Answer A is wrong. Caster may be adjusted by adding or removing shims.
Answer B is wrong. Caster may be adjusted by lengthening or shortening the strut rod.
Answer C is correct. Torsion bar adjusting bolt is used to adjust ride height.
Answer D is wrong. Caster may be adjusted by turning the eccentric cams.

Glossary

Adjusting sleeve An internally threaded sleeve located between the tie-rod ends that, when turned, moves the front or rear wheels when toe is being adjusted.

Aftermarket A term used for parts and equipment sold by companies independent from the vehicle manufacturer.

Air bag A passive restraint system having an inflatable bladder located in the steering wheel and in the dash ahead of the passenger seat.

Air shock A shock using air pressure and a rubberized bag that is a part of the shock absorber.

Air shock system Also known as a secondary air leveling/suspension system. Uses air pressure and a rubberized bag that is a part of the shock absorber.

Air suspension system A suspension system that uses compressed air and air springs to replace conventional coil springs.

Alignment The act of lining up; of being in a true line.

Alignment rack A drive-on device used for alignments.

Aspect ratio The relationship between the height of the tire and the tread width.

Axial play Movement of a component parallel to its axis.

Axle A cross member supporting a vehicle on which one or more wheels are mounted.

Axle flange The outside end of the axle where the wheel/drum/rotor attaches.

Ball joint A joint or connection where a ball rides in a socket.

Battery A device that stores electrical energy in chemical form.

Bearing An antifriction device having an inner and outer race with one or more rows of steel balls.

Belt A device used to drive the water pump and other accessories.

Boot A flexible rubber or plastic cover used over a component to protect it from the elements.

Bump steer When a vehicle experiences a toe change over a bump, it is called bump steer. Almost all vehicles have some bump steer.

Bushing A sleeve, usually bronze, inserted into a bore to support a shaft.

Camber The inward or outward tilt of the wheel and tire assembly as viewed from the front or rear of the vehicle.

Caster The angle between the steering/spindle axis and vertical, as viewed from the side of the vehicle.

Center bolt A term used for a bolt that holds the spring leaves in a fixed position and indexes the spring to the axle housing.

Center link A steering linkage part that is connected to the tie-rod ends and the Pitman arm to transfer the rotating motion of the steering box to linear motion necessary to move the wheels from side to side.

Chassis The frame of a vehicle.

Clock spring Device attached to the steering column and the steering wheel that insures good electrical paths for the air bag assembly signals.

Coil spring A spring steel bar or rod that is shaped into a coil to provide an up-and-down springing effect.

Conicity A tire condition that occurs when the tire tread is installed off center on the carcass, creating a cone shape. This causes the vehicle to pull to one side and is diagnosed by changing the tire position on the vehicle.

Connecting link A link in which a removable rod or plate facilitates connecting or disconnecting the ends of a chain.

Control arm The main link between the vehicle wheels and frame.

Crash sensor Normally open input of the air bag system that has gold plated contacts that close when subjected to a predetermined deceleration force.

Directional tires Tires with a particular tread pattern that are designed to give maximum traction. Directional tires must be mounted to turn in a specific direction of rotation.

Drivetrain All of the components required to deliver engine power to the road surface.

Dynamic Pertaining to energy, force, or motion in relation to force. A term used when wheels are balanced using equipment that spins the tire and wheel assemblies.

Eccentric adjuster An adjustment system that converts a rotary motion into a reciprocating motion.

Engine The prime mover of a vehicle that converts chemical energy (fuel) into mechanical energy (motion).

Exhaust The harmful burned and unburned gases that remain after combustion; the pipe extending from the muffler to vent the gases.

Exhaust system The exhaust manifold, catalytic converter, muffler, and pipes that vent harmful burned and unburned gases, a product of combustion, to the atmosphere.

Factory bulletin An official periodic publication by the vehicle manufacturer with service tips and hints.

Flush To use a fluid to remove solid and semisolid particles.

Frame The substructure of a vehicle.

Free play The movement permitted between two mating and/or rotating parts.

Front cradle Heavy metal framework that attaches to the chassis and supports the powertrain and allows the attachment of many of the suspension and steering parts.

Front-wheel drive A drive system that transmits power through the front wheels.

Full assist A term used with variable assist steering systems to signify normal power system operation without restrictions.

Fuse A safety device to guard against an electrical overload.

Grease A heavy lubricant.

Height sensor A sensor used to monitor the corner height of a vehicle for the level control system.

Hose A rubber or synthetic tube used to transfer fluid or vapor.

Hub The center component of a wheel or gear.

Hub runout The amount of wobble out of plane measured at an outside machined surface of the hub.

Idler arm A pivoting component that attaches to the side of the vehicle opposite the Pitman arm and supports the center link, allowing parallel movement of the steering linkage.

Idle relearn The routine of the computer that resets idle requirements when a disconnected battery has been reconnected.

Included angle The sum of steering axis inclination and camber.

Intermittent problem A problem that only occurs occasionally.

Isolator pad A rubber or synthetic device used to insulate two parts to reduce noise and/or vibration.

Km/hr An abbreviation for kilometers per hour; one km/hr is equal to 0.62 mph.

Knuckle The part around which each front wheel pivots as it is steered.

Leak spring A vehicle suspension component having one or more thin steel leaves.

Load carrying A term often used when referring to ball joints. It could be the upper or lower ball joint depending on where the spring supporting the vehicle weight is located.

Load rating A term used to designate the maximum weight a tire is designed to support; the maximum weight a vehicle is designed to support and/or carry.

Lubricant A fluid used to reduce friction between moving parts.

Lubricate To provide lubricant to areas that require such service.

MacPherson strut An independent suspension part that, in connection with the coil spring that supports the vehicle, serves as a shock absorber. If on the front of the vehicle, the spindle, ball joint, and steering arm are attached.

Memory (memory code) A term used for certain fault codes stored in the module of an electronically controlled computer that happened during past operation but is not now present as a hard fault.

Module A solid-state device composed of the components required to monitor, control, and sometimes diagnose many of the systems of today's vehicles.

MPH (mph) An abbreviation for miles per hour.

Neutralized A term used for allowing a system to settle in a nonstressed position. Used in reference to exhaust systems, powertrain mounts, and suspension mounts.

Non-load carrying A term used in reference to the ball joints on a vehicle that hold the suspension parts in place, but do not support the vehicle load.

Oil A mineral based lubricant.

O-ring Round or square donut-shaped rubber or synthetic device used to seal a joint or shaft.

Pitman A term often used for the Pitman arm and Pitman shaft.

Pitman arm An arm attaching the steering box sector shaft to the center link that changes the rotating forces of the steering box to linear motion and moves the front wheels from side to side.

Power steering A power assisted steering system.

Power steering analyzer A diagnostic device that tests flow, pressure, and operation of everything hydraulic in the power steering system.

Power steering fluid A special fluid used in power steering systems.

Power steering pump A component of the power steering system providing fluid power for operation.

Power supply An electrical or electronic device that provides a predetermined power to satisfy a particular requirement.

Pre-alignment check A series of checks the technician must perform before an alignment is done. All steering and suspension parts are checked for looseness, wear, damage, and general condition.

Preload The specified pressure applied to certain parts during assembly or installation.

Pressure A force per unit area, usually measured in pounds per square inch (psi) or kilopascals (kPa).

Pressure control valve A device used to regulate and/or control a pressure.

Primary air suspension system The term used for a type of vehicle system that uses air, instead of coil, leaf, or torsion bar type springs, to support the vehicle weight.

Pulley A wheel-shaped belt-drive or belt-driven device.

Rack A horizontal toothed bar in the rack-and-pinion power steering sector.

Rack-and-pinion system A type of steering assembly having a pinion on one end that engages in a horizontal-toothed rack with tie rods at either end that attaches to the steering arms.

Rack body A part of the steering system containing the rack.

Rack guide The guide on which the rack is positioned.

Rack shaft A horizontal-toothed part of the rack and pinion.

Radius A term used when referring to the turning radius of a vehicle; a line extending from the center of a circle to its boundary.

Rear end The differential and final drive assembly on a rear-wheel-drive vehicle.

Rear springs Coil- or leaf-type suspension components at the rear of a vehicle.

Rear strut A shock-absorber-type component that supports the rear of the vehicle.

Redundant A duplication; a secondary or backup system.

Relay An electromagnetic/mechanical switch.

Road test To drive a vehicle and determine needed repairs.

Rpm An abbreviation for revolutions per minute.

Sector shaft A shaft on which the sector gear is located. A component of a conventional steering gear.

Self test A computer function that scans subsystems and systems to provide data for troubleshooting.

Sensor An electronic device used to monitor relative conditions for computer control requirements.

Shackle bushing Insulated bushings to help prevent the transfer of noise and road shock.

Shim A thin metal spacer used to align two parts.

Shimmy Harsh side-to-side vibration of the steering transmitted to the steering wheel; usually caused by loose suspension parts or front wheel imbalance.

Shock absorber A hydraulic cylinder located at each wheel of the suspension system to dampen road shock.

Silencer pad A rubber or synthetic device used to insulate two parts to reduce noise and/or vibration.

Solenoid An electro-mechanical device used to impart a push-pull motion.

Speed rating A tire rating that indicates the maximum safe speed the tire is designed for.

Speed sensor An electrical device that can sense the rotational speed of a shaft and transmit that information to another device, such as a computer.

Spindle A shaft, stub axle, or knuckle upon which wheel hubs and wheel bearings ride.

Spring A coil- or leaf-type device used in the suspension system for springing the vehicle weight.

Sprung weight The entire weight of the vehicle components that are supported by the springs.

Stabilizer bar link Long metal bolts with insulators and support washers that attach the stabilizer bar to the outer suspension or body.

Static A commonly used term for a balancing process that compensates for an imbalance condition by distributing the weight equally around a part without rotating the part being balanced.

Steering The method whereby the vehicle is kept on course.

Steering arm An arm attached to the steering knuckle/spindle/strut that moves those parts in response to steering wheel movements.

Steering axis inclination (SAI) The angle of a line through the center of the upper strut mount and lower ball joint in relation to the true vertical centerline of the tire, as viewed from the front of the vehicle.

Steering box A general term used for the steering mechanism at the end of the steering column.

Steering column The tubing through which the steering shaft mounts and rotates.

Steering damper A device that reduces or eliminates road shock and vibration.

Steering knuckle The part which pivots in response to forces from the steering box or rack-and-pinion gear and causes the tires to hold the vehicle on course and control the direction of movement of the vehicle.

Steering linkage The assembly of the tie rods, idler arms, and links that make up the system that transfers steering motion to the front wheels.

Steering linkage damper A shock-absorber-type device that connects the steering linkage to the framework of the vehicle and absorbs some of the road shock and dampens most of the vibrations. Such devices are used on many trucks and off-road vehicles.

Steering wheel The wheel located at the top of the steering shaft that the driver uses to steer the vehicle.

Steering wonder The tendency of a vehicle to pull to one side when driven straight ahead.

Strut A component, connected at the top of the steering knuckle to the upper strut mount, that maintains the knuckle position.

Subframe A partial front or rear chassis frame used on some vehicles to support the powertrain and suspension and steering assemblies.

Suspension The system that supports the weight of the vehicle and provides a comfortable and safe environment for the occupants.

Suspension height The height of the vehicle at its four corners.

Sway bar A bar in the suspension system that connects the two sides together in a manner that cornering forces or a road shock is shared by both wheels.

Technical service bulletin Bulletins provided by the manufacturer regarding production changes and corrections to aid the technician in troubleshooting practices and procedures.

Tie rod The steering linkages between the idler arm, Pitman arm, and steering arm.

Tie-rod end A pivoting ball-and-socket joint located at one end of the tie rod.

Tire information label A label required by the federal government and usually placed on the inner glove box door or on a door post on the passenger side of the vehicle. It contains all of the information needed to chose a safe tire of the proper size for the vehicle.

Tire wear pattern The way in which a tire wears due to front end or balance problems.

Toe A suspension dimension; the difference between the extreme front and extreme rear of a tire.

Toe adjustment The methods provided by the manufacturer to move the front or rear wheels so there is a specified distance between a centerline of each front wheel, and the centerline of the rear wheels.

Torque The measure of a force producing tension and rotation around an axis.

Torsion bar A long, spring-steel bar that is used instead of a coil or leaf spring. It usually has an adjustment at one end that makes it possible to easily adjust suspension height.

Transmission A manual or automatic device; a part of the drive train that provides different input and output ratios.

Unscheduled deployment A term used when referring to air bag systems that could be set off without the vehicle being involved in a crash severe enough for normal deployment. Also, accidental deployment of the air bag by not following the vehicle manufacturer's safety precautions.

Unsprung weight The components of a vehicle that rest directly on the road which are not supported by the springs. Wheels, tires, differentials and half of the weight of any component like control arms that bolt to the wheel and body are the unsprung weight.

Valve stem A device found in the rim of a tire to provide a means of adding air and/or checking its pressure.

Vehicle A means of conveyance or transportation.

Vehicle ride height The specified normal distance between the vehicle chassis and a level surface.

Wheel A circular frame or hub of an axle to which a tire is mounted.

Wheel balance The equal distribution of the weight of a wheel with a mounted tire.

Worm gear A component of a steering gear, into which teeth are cut, resembling the threads of a screw.

Worm shaft A steering gearbox component having spiral grooves.

Notes

Notes

Notes

Notes

Notes

Notes

Notes

Notes

Notes